DC SERVOS

Application and Design with MATLAB®

DC SERVOS

Application and Design with MATLAB®

Stephen M. Tobin

CRC Press
Taylor & Francis Group
Boca Raton London New York

CRC Press is an imprint of the
Taylor & Francis Group, an **informa** business

MATLAB® and Simulink® are trademarks of the Math Works, Inc. and are used with permission. The Mathworks does not warrant the accuracy of the text or exercises in this book. This book's use or discussion of MATLAB® and Simulink® software or related products does not constitute endorsement or sponsorship by the Math Works of a particular pedagogical approach or particular use of the MATLAB® and Simulink® software.

CRC Press
Taylor & Francis Group
6000 Broken Sound Parkway NW, Suite 300
Boca Raton, FL 33487-2742

First issued in paperback 2017

© 2011 by Taylor and Francis Group, LLC
CRC Press is an imprint of Taylor & Francis Group, an Informa business

No claim to original U.S. Government works

ISBN-13: 978-1-4200-8003-2 (hbk)
ISBN-13: 978-1-138-11385-5 (pbk)

Library of Congress Cataloging-in-Publication Data

Tobin, Stephen M.
    DC servos : application and design with MATLAB / author, Stephen M. Tobin.
        p. cm.
    "A CRC title."
    Includes bibliographical references and index.
    ISBN 978-1-4200-8003-2 (hardcover : alk. paper)
    1. Servomechanisms--Design and construction. 2. Electric controllers--Design and construction. 3. Automatic control--Data processing. 4. MATLAB. I. Title.

TJ214.T63 2011
629.8'323--dc22   2010028304

Visit the Taylor & Francis Web site at
http://www.taylorandfrancis.com

and the CRC Press Web site at
http://www.crcpress.com

Dedication

To Elisa

Contents

Preface

The idea for this book came from years of frustration as I tried to find an authoritative source for how to design a custom servomechanism, using a few reference designs with detailed explanations and electrical schematics that can be readily built, tested, and changed if necessary by the reader. These designs would use modern, easily accessible parts. The primary focus would be on the use of commonly available brushed DC motors in position and velocity control systems. There was a particular void in the literature regarding the application of optical encoders. Many texts would describe these feedback transducers in some detail, yet fail to show how they are used in a real system. In this new book, there would be an even treatment of the theoretical and practical sides of application and design, seemingly unavailable to the author, even though this work has been going on for more than 50 years.

When I graduated from the University of New Hampshire in 1983, robots were doing real work inside automobile manufacturing plants. During our senior year, a few fellow student friends and I decided to build a small robot to see what was going on with all this "motion control" stuff. We were hooked, and our senior project took two semesters to complete. Early on in the project, we decided that we would use step motors for the prime movers in our robot. The reason was that our advisor for the project, Charles K. ("Charlie") Taft, was a respected authority in the modeling and use of steppers. Charlie had friends in industry who could supply us with free motors. It was simply the quickest means to our end, and even though three of us had taken and passed EE/ME 781 (the UNH Control Systems course), we had no idea how to implement a DC servo. At the time, the course was too theoretical for us to absorb. We had a gut feeling that DC servos would give us a more flexible, robust solution but did not know why. Charlie simply did not have the time or the energy to teach us. He was in high demand in the UNH ME department and had several graduate students to attend to, in addition to his regular undergraduate course load. In the end, we created a step motor–controlled robot that had very limited capabilities (see Figure P.1).

Despite our relative inexperience as students in 1982, I continue to believe our instincts were correct. Motion control mattered in the real world. This is borne out by a prediction made by Caxton C. Foster in his 1981 book *Real Time Programming—Neglected Topics* (see bibliography). In the book, he explored subjects that defied classification into engineering curriculums at that time. Nominally, his audience was computer science majors

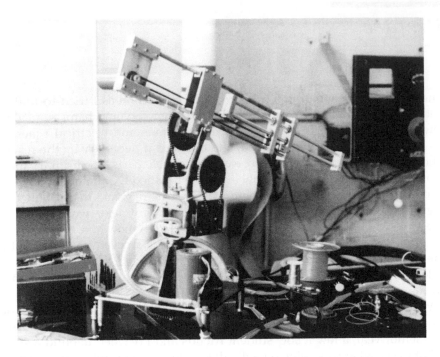

FIGURE P.1
A step motor–controlled robot, designed and built in 1983 at UNH by the author and three other collaborators.

and computer hobbyists. One relevant question he posed was "Why study servomechanisms?" His answer still holds true:

> With the dramatic decline in the price of computers over the past ten years, they have moved out of the accounting departments of business and the ivy-covered walls of academia into private homes and industrial assembly lines. One of the major applications of the new micros has been, and will continue to be, in controlling physical devices of one sort or another. Now the fundamental idea behind a servomechanism is the application of feedback to controlling position or velocity or some other attribute of the device in question.

Foster goes on to give an eloquent, concise introduction to the fundamental aspects of control and what role the new digital computers could play in it. In retrospect, it would have been great to have taken a course like his as a mechanical engineering major, or at least be exposed to his book at the time. However, I truly believe that learning is a lifetime endeavor. I admit it—I still get a rush every time I see the motion of an object being controlled. That is why I wrote this book.

Today, DC servos are working all over the world in countless applications—CD players, ink-jet printers, robotics, machining centers, vending machines, eyeglass manufacturing, home appliances, automotive seat positioning. How are all these machines being designed? In my experience working at a number of small to mid-sized companies, there are one or two key people who have been through "servo school." How did they get there? What was their path to learning all the nuances of servo design? How did they find out how to implement these design techniques in a real-world product? The fact is that successful servo control of mechanisms draws from so many different engineering disciplines that it is difficult for anyone to master it all. It takes years of experience (for most of us) to grasp all of the details necessary. This book is an attempt to distill some, if not most, of the practical knowledge required from the various branches of applied science—electrical engineering, analog electronics, mechanical engineering, mechanics, control theory, digital electronics, embedded computing, and firmware design—into a cohesive framework. The goal of this book is to take you on a short journey through servo school and to teach you what I had to learn on my own. I feel compelled to do this, so that more engineers, students, scientists, and hobbyists can feel confident designing the ever more complex machines of the 21st century. With this book, I hope both author and reader will make meaningful contributions to society at large.

Stephen M. Tobin
Newton Highlands, Massachusetts

Acknowledgments

I wish to thank my publisher, Taylor & Francis, as well my editor, Jonathan Plant, project coordinator, Amber Donley, and project editor, Andrea Demby of CRC Press. They stuck with me even though this project took twice as long to complete as first anticipated. I also wish to thank Judy Bass at McGraw-Hill, who patiently taught this first-time author how to promote a work such as this one within the publishing community. Diane Trevino at National Semiconductor was very generous with permission to use and reproduce several application notes. Harprit Sandhu (along with his editor at McGraw-Hill, Roger Stewart) was extremely generous with his sharing of firmware, without which Chapter 8 could not have been written in such a reasonable length of time. I also wish to thank The MathWorks and their book program staff for complimentary copies of MATLAB® and Control System Toolbox software used throughout the book.

Some personal associations and friendships must be duly noted. My illustrator, Bob Conroy of Mantis Design Associates, has become a trusted collaborator and guide throughout the entire project. Charles K. ("Charlie") Taft was my first great teacher, and truly inspired me while an undergraduate at UNH. One of my senior project partners at UNH, Brian Butler, continues to be a great friend and source of wisdom. His statement to me some years ago, "Your best is good enough," has helped sustain me throughout my adult life. Samuel Fuller and Michael Hunter were my other two partners on the UNH robot project. Finally, I wish to thank my wife, Elisa, for her unwavering support over the past two years, making this book the best it could possibly be.

Since no book, human being, or DC servo system is perfect, errors and omissions are unavoidable. Your feedback would be very much appreciated. Please forward your comments to me directly at optical-tools@comcast.net.

About the Author

Stephen M. Tobin is the founder and president of Optical Tools Corporation. He received his B.S. degree in mechanical engineering from the University of New Hampshire in 1983. He spent the first five years of his career in the motion control field, holding positions in step motor and optical encoder design engineering with divisions of Allied-Signal and Dresser Industries. He then turned his attention to the development of electro-optical instrumentation. He joined General Eastern Instruments (now a division of General Electric) in 1988, working on closed-loop optical humidity measurement systems. He received his M.S. degree in electrical engineering with a concentration in electro-optics from Tufts University in 1994. He later spent six years developing medical devices at Arthur D. Little, a world-renowned consulting firm based in Cambridge, MA. Optical Tools Corporation (http://www.optical-tools.net) was founded by Tobin in 2004, and he continues to consult to the medical device and manufacturing automation communities. He holds four U.S. patents and is a member of the Tau Beta Pi National Engineering Honor Society. Motion control continues to be his life long passion.

1

DC Servo Systems Defined

1.1 Scope and Definition

In the roughly 40 years between man landing on the moon and today, the DC servo has been implemented in countless applications in industry. The invention of solid-state amplifiers, as well as the microprocessor, has given rise to the idea of preprogrammed machines running on direct current. The more recent resurgence of smaller DC servos running on less than 100 watts can be attributed to a few factors: (1) the revolution of personal computers and the need for their "peripheral" electromechanical devices; (2) market demand for miniaturization and portability of increasingly complex machines; and (3) recent worldwide concern for electrical safety, where governmental approvals are much easier to obtain when a device runs on less than 48 volts DC. A typical example is the print head positioning relative to the paper on an ink-jet printer. *Servo* is derived from the Latin noun *servus*, which means "slave." For now, we define the servo as a machine that can perform a predefined action of larger power output than that exerted by the operator.

1.2 The Concept of Feedback Control

Consider the simple act of boiling water for a morning cup of tea. A modern automatic teakettle uses a temperature sensor to detect the onset of boiling and turns itself off. In Figure 1.1, an electrical signal representing the water temperature is constantly measured with respect to a reference value (i.e., fed back), and an action is taken upon reaching that value. Contrast this with the standard whistling kettle that requires human intervention to hear the whistle, walk to the cooktop, and shut down the power source.

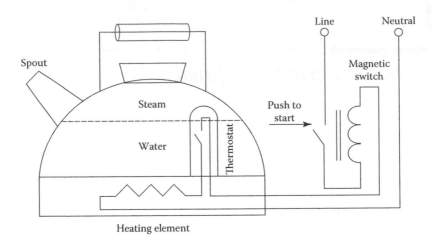

FIGURE 1.1
Electric teakettle control schematic.

1.3 Types of Control

We further classify the natural control types as (1) open loop vs. closed loop control and (2) on/off vs. continuous control. Brief descriptions of these follow, using the teakettle as an example.

1.3.1 Open Loop vs. Closed Loop Control

As described above, if the water is boiled with human intervention, we refer to this as an "open loop" process, since no feedback is employed. Conversely, if the kettle is automatically shut off after the boiling task is completed, this process is a "closed loop" one.

1.3.2 On/Off vs. Continuous Control

The automatic teakettle is a simple on/off or "bang-bang" control system. Power is either applied fully or not applied at all. In a continuous control process, the amount of power applied can be dependent on the magnitude of the feedback signal, its sign, its rate of change (derivative), or its accumulation (integral), along with various combinations of these.

1.4 Comments on Motion Control

As considered in this book, the DC servo is fundamentally concerned with the concept of motion, either translational or rotational. The motion to be controlled will usually be the position, or its first derivative, velocity of a

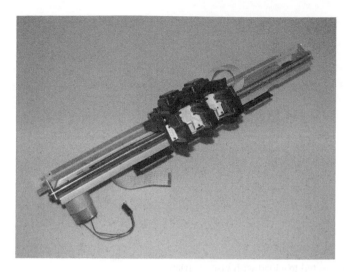

FIGURE 1.2
Ink-jet printer cartridge transport subsystem. Note the transparent linear encoder strip, bounded by black dots on either side and suspended in tension between the assembly end supports.

given load. DC servo control is usually of the closed-loop, continuous variety, unless a step motor is used as a prime mover. We will not consider step motors here, but the reader may consult references at the end of the book for further information.

1.4.1 Continuous-Time vs. Discrete-Time Motion Control

In the past 20 years, inexpensive microcontrollers have given designers exciting new options in control systems. An excellent introduction to how computers can be interfaced with DC servos is given by [Foster, 1981]. Especially in high-volume applications, competition drives more features, lower weight, and lower cost into products such as ink-jet printers. All-digital implementations of the DC servo have become the norm in these products to keep production costs down. Usually pulse-width-modulated amplifiers taking a digital input are used with incremental encoders that provide pulsed output as the feedback mechanism. We refer to these "all-digital" designs as discrete-time servos. A photograph of an ink-jet printer cartridge transport subsystem implemented as a discrete-time servo is shown in Figure 1.2.

1.5 Introduction to a DC Motor Driving a Mechanical Load

The following analysis is mainly due to Doebelin [Doebelin, 1972]. Figure 1.3(a) we will call a functional schematic, essentially a sketch of the system. We are interested in the response of the load speed ω_o to an input

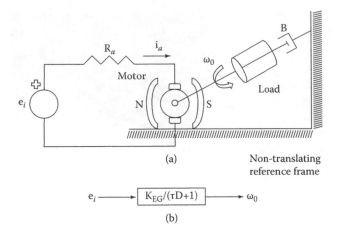

FIGURE 1.3
DC motor connected to a load with viscous friction.

voltage e_i. We assume the "field" of the motor is supplied by permanent magnets, which is valid in virtually all cases today with the development of high coercive force magnetic technology. We also will ignore the armature inductance for this initial study. We can do this because in dynamic systems, we are usually interested in the components that have the slowest reaction to a given input change. In a mechatronic system such as this one, solid bodies like a motor's rotor and attached load have a slower reaction time, in general, than the time taken to build up a magnetic field in the inductance of the armature of a small motor. With this in mind, we can write for the electrical circuit

$$i_a R_a + K_s \omega_o = e_i \tag{1.1}$$

where

$i_a \triangleq$ *armature current, amp*

$R_a \triangleq$ *armature resistance, ohm*

$K_e \omega_o \triangleq$ *voltage drop due to back emf of motor, volt*

$K_e \triangleq$ *motor back emf constant,* $\dfrac{volt}{\frac{rad}{sec}}$

Next applying Newton's law to the motor shaft, we have

$$K_T i_a - B \omega_o = J \dot{\omega}_o \tag{1.2}$$

where

$K_T \triangleq$ *motor torque constant, Newton · meter/amp*

$B \triangleq$ *combined viscous damping of motor and load, N · m · sec*

$J \triangleq$ *combined inertia of motor and load, N · m · sec²*

In Equation (1.1), the back emf voltage must oppose the voltage supplied by e_i. Similarly, the damping term of Equation (1.2) opposes the torque produced by the applied armature current.

Equations (1.1) and (1.2) constitute a model of the physical system shown in Figure 1.3(a). Given two equations and two unknowns, we can eliminate the armature current and solve for the load speed. From (1.1) we have

$$i_a = (e_i - K_e \omega_o)/R_a \tag{1.3}$$

From (1.2) we can place the time derivative of the load speed in terms of the speed itself algebraically by invoking the D-operator notation used by [Reswick and Taft, 1967], [Doebelin, 1962], and others. Rewriting (1.2) we have

$$K_T i_a - B \omega_o = JD \omega_o \tag{1.4}$$

where

$$D \triangleq \frac{d}{dt} \tag{1.5}$$

Considering the load speed as the system output and the applied voltage as the input, we can solve for the system transfer function, the quotient of output and input, in simplified form. Substituting (1.3) into (1.4) we get

$$\frac{\omega_o}{e_i} = K_{EG}/(\tau D + 1) \tag{1.6}$$

where

$$\tau \triangleq \textit{system time constant, sec} = J/\left(B + \frac{K_T K_E}{R_a} \right) \tag{1.7}$$

$$K_{EG} \triangleq \textit{system gain,}\left(\frac{r}{s} \right)/V = K_T/(BR_a + K_T K_E) \tag{1.8}$$

Transfer functions are conveniently illustrated by block diagrams, and the simple transfer function of (1.6) is shown in Figure 1.3(b). In dynamics, this system is considered "first-order," because it involves only a first derivative in its differential equations (1.1) and (1.2). It has a steady-state gain and a time constant. The dynamics of first-order systems are well covered many references, including [Cannon, 1967] and [Evans, 1954]. Doebelin also points out that the quantity $K_T K_E / R_a$ has the same numerical units as B and causes a viscous damping effect. So, even if the damper B was not included in our system model, damping is still present. This system's parameters are dependent on both electrical and mechanical quantities, and it is indeed possible to merge these quantities from seemingly different "branches" of engineering into a cohesive framework. It is the author's experience that successful engineers bridge this gap between "electrical" and "mechanical" labels with an uncommon but attainable ease. Readers are encouraged to allow this gap to fade as much as possible.

1.6 Realization of a Velocity Servo

If we were to "turn on" the system of Figure 1.3(a), we would see that a particular setting of the input voltage e_i might regulate the load speed to within 10 percent. Why? For example, as the motor windings heat up their resistance will change. The field strength of the magnets is also affected by the rising temperature of the nearby windings. Suppose we need more precise speed control in our application, say within 1 percent. Let us try the following to accomplish this:

1. Measure what the speed actually is.
2. Compare this value to a desired value.
3. Adjust the voltage e_i to reduce the error if one exists.

This procedure, illustrated in Figure 1.4, is known as feedback—the cornerstone of all control work. Practically, we need to couple a tachometer to a rear shaft extension on the motor, giving a voltage proportional to the actual speed. We then add a control voltage representing the desired speed to a closed circuit that feeds an error voltage, if any, to an electronic amplifier. Note that the negative terminal of the tachometer is connected to the negative terminal of the control voltage source, so that the difference between the two is fed to the amplifier input. The output of this amplifier becomes our new input voltage to the motor. A block diagram of this setup is shown in Figure 1.5(a). The system has become more complex, but the benefits can far outweigh the costs as we shall soon see.

To clearly illustrate the value of feedback, we shall employ block diagram algebra [Mancini, 2003; DiStefano, 1967]. Since values of speed in this control

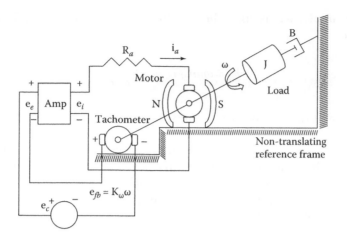

FIGURE 1.4
DC motor speed control using feedback.

system are translated to voltages, we will scale the control setting e_c by the sensitivity of the tachometer:

$$\omega_{o,des} \triangleq desired\ speed, \frac{rad}{sec} = \frac{e_c}{K_\omega} \tag{1.9}$$

where

$$e_c \triangleq control\ voltage,\ volts \tag{1.10}$$

$$K_\omega \triangleq tachometer\ constant, \frac{volts}{\frac{rad}{sec}} \tag{1.11}$$

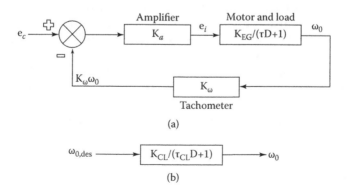

FIGURE 1.5
DC motor speed control block diagrams.

In this example, we assume the amplifier to be a constant gain element, such that

$$K_a \triangleq amplifier\ gain, \frac{volt}{volt} \tag{1.12}$$

(This assumption is valid, since solid-state amplifiers usually "roll off" in the range of tens or hundreds of kilohertz. They are essentially low-pass filters with a pass band much wider than that of a DC servo.) The benefits of feedback can become clear if, for the purposes of this example, we reduce the multi-block diagram of Figure 1.5(a) to a single block similar to Figure 1.3(b). Using algebraic manipulation, the following two equivalent closed-loop system characteristics can be derived:

$$\tau_{cl} \triangleq closed\ loop\ time\ constant,\ sec = \frac{\tau}{1 + K_{EG}K_aK_\omega} \tag{1.13}$$

$$K_{cl} \triangleq closed\ loop\ speed\ gain, \frac{\frac{rad}{sec}}{\frac{rad}{sec}} = 1 / \left[1 + \left(\frac{1}{K_{EG}K_aK_\omega} \right) \right] \tag{1.14}$$

The single block equivalent system is shown in Figure 1.5(b). It is usually not necessary to reduce systems like this to one block, but it is being done here to easily compare the characteristics of the open-loop system to those of the closed-loop system. Although both systems are first order, the results are profound if we examine Equations (1.13) and (1.14) in more detail. The quantity $K_{EG}K_aK_\omega$ is commonly called the "loop gain," because it is the product of the individual gains going around the loop of Figure 1.5(a). If this quantity is made large relative to 1.0, the system performance improves dramatically, even though the motor and load have not changed. Doebelin warns that although it is easy to do in practice, we must take care not to raise the gain too much, for instability may result. The model presented here does not predict this instability—we have ignored certain system elements that become important as the loop gain is raised. As a simple example, suppose K_{EG}, K_a, and K_ω are all equal to 1.0. If we raise the amplifier gain K_a to 10.0, the system time constant becomes $\tau/11$, making the feedback system 11 times faster. Similarly, the closed-loop speed gain, which relates the desired speed to the actual speed in the steady state (after transient conditions have disappeared), is

$$\frac{\omega_o}{\omega_{o,des}} \rightarrow SS \triangleq K_{cl} = 1 / \left(1 + \frac{1}{\left[\frac{K_T}{BR_a + K_T K_E} \right](K_a K_\omega)} \right) \tag{1.15}$$

We will examine this idea of a steady-state error in more detail later. If we follow Doebelin's example, suppose K_T, B, R_a, and K_E are all equal to 1.0, and $K_a K_\omega = 40$. In this case,

$$\frac{\omega_o}{\omega_{o,des}} \rightarrow SS = \frac{1}{1+\frac{1}{20}} = 0.952 \tag{1.16}$$

Now suppose that heating of the motor, due to winding heating or external effects, causes the permanent magnet field to change, in turn causing K_T to drift to a value of 2.0. (R_a would also be affected in this case, but we will ignore this for the sake of simplicity.) We now have

$$\frac{\omega_o}{\omega_{o,des}} \rightarrow SS = \frac{1}{1+\frac{3}{80}} = 0.964 \tag{1.17}$$

Thus a 100 percent change in K_T has caused only about a 1 percent change in speed, which is in the range of speed control we originally wanted. Without the feedback system employed, from (1.8) we would have had a change from

$$K_{EG1} = \frac{K_T}{BR_a + K_T K_E} = \frac{1}{1+1} = 0.5 \tag{1.18}$$

to

$$K_{EG2} = \frac{K_T}{BR_a + K_T K_E} = \frac{2}{1+2} = 0.667 \tag{1.19}$$

This represents a 33 percent speed change. The use of feedback has allowed us to achieve much tighter control of the load speed, in spite of effects that would have been very difficult to compensate for otherwise. This powerful method is used not only for controlling motion but also for countless other processes in science and engineering.

References

Cannon, R., *Dynamics of Physical Systems*, McGraw-Hill, 1967.
DiStefano, J. et al., *Feedback and Control Systems*, Schaum, 1967.
Doebelin, E., *Dynamic Analysis and Feedback Control*, McGraw-Hill, 1962.
Doebelin, E., *System Dynamics: Modeling and Response*, Merrill, 1972.
Evans, W., *Control System Dynamics*, McGraw-Hill, 1954.
Foster, C., *Real Time Programming—Neglected Topics*, Addison-Wesley, 1981.
Mancini, R., *Op Amps for Everyone*, Newnes, 2003.
Reswick, J., and C. Taft, *Introduction to Dynamic Systems*, Prentice-Hall, 1967.

2

Anatomy of a Continuous-Time DC Servo

2.1 Description

The follow-up shaft repeater has been described in numerous texts covering system dynamics and servomechanism theory. (A servomechanism is simply a servo system whose output is a mechanical motion.) Figure 2.1 shows a functional schematic of a shaft repeater, where the output shaft of the gear train is caused to follow the motion of the input potentiometer. Its widest application has been defense-related hardware positioning since World War II. For example, the rudder position on a large sea vessel could have been positioned from the command deck via a shaft repeater. From Figure 2.2, it is seen that the rudder will follow the wheel position if suitable amplification is used.

2.2 Intended Use

The shaft repeater to be presented in this chapter is an all-analog implementation with a range of motion less than 360 mechanical degrees. Although the utility of this servomechanism may seem limited, we can argue its relevance in a few ways. First, it has tremendous value as a teaching tool because it is compact and relatively inexpensive. Second, we will use human input to activate it, which is useful for demonstrating the value of experience to the reader. Third, it has applications in military hardware, as well as in automatic volume control of sound systems and remote spotlight positioning, just to name a few. Most DC servo systems today are commanded by preprogrammed electrical signals, one of the most prominent being computer numerical controlled ("CNC") centers for automated machining of metallic parts. In fact, the shaft repeater presented here is commanded by an electrical signal; it just so happens that this signal is being generated by a human

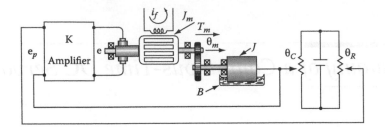

FIGURE 2.1
DC servo shaft repeater schematic.

being turning a potentiometer. For the curious reader, a hybrid analog/digital multi-turn servomechanism, controlled by a microprocessor, is given in [Foster, 1981]. Foster's device, as a learning tool, is an interesting and valid substitute for the servo to be presented here.

In this book, we will use the follow-up shaft repeater in an unusual way relative to most textbooks. It will first be introduced in its entirety, with a full electrical schematic and parts list, making the system immediately real to the reader. It will be used as a foundation for future chapters of the book, and we will go back and review it as needed to reinforce important concepts.

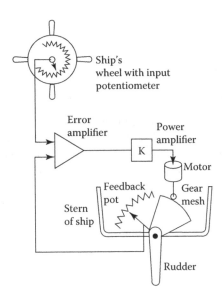

FIGURE 2.2
Example of a ship's rudder control.

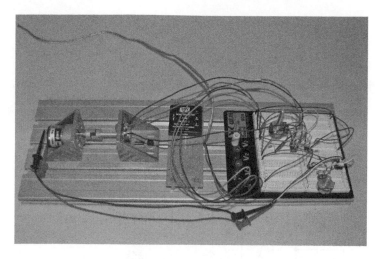

FIGURE 2.3
The shaft repeater prototype system.

2.3 The Prototype

A shaft repeater as given in Figure 2.1 has been built and tested for the purposes of this book. Figure 2.3 shows a photograph of the system, and Figure 2.4 gives a complete electrical schematic. A few factors regarding its development are noted. The system functions without the use of a compensation network (denoted by op-amp U3A and its surrounding components) and some intermediate gain (denoted by op-amp U3B and its components). Without these, the servo is a "proportional controller," which will be explained later in the book. With one or both of these in place, performance can be improved. This is commonly referred to as "compensating" the system, and this will also be the subject of a later chapter.

2.4 Electrical Design and Construction

As shown in Figure 2.3, the shaft repeater was initially designed without the use of op-amps U3A and U3B. The system functioned in a stable manner, but we anticipate that some sort of compensation will be required to get acceptable response. The reasons for why this type of compensation was chosen will become apparent later. In a design environment, usually one starts with

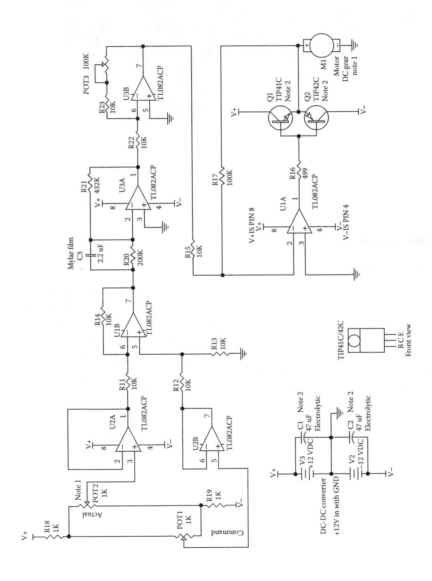

FIGURE 2.4

Shaft repeater system electrical schematic. Note 1. POT2 is mechanically coupled to M1. Note 2. C1 and C2 to be placed physically close to Q1 and Q2, respectively.

an initial design based on best engineering judgment. As [Doebelin, 1962] suggested more than 40 years ago:

> In preliminary studies, usually very gross assumptions are made in order to get a quick overall picture with [relatively] little expenditure of time and effort. Later, when the situation has jelled to some extent, more precise analyses are in order. Intelligent choice of simplifying assumptions depends upon clear knowledge and proper application of fundamental principles and also upon experience with real systems and their actual experimentally measured behavior.

We emphasize here that this is the kernel of the engineering development process, and we will be following the same path in this book. A clear example of this process in action was the inclusion of the compensation scheme and its component selection. The author's experience says that including a lead network for compensation is desirable. Typical values for the attributes of a DC motor were taken from the example used by The MathWorks in their Control System Toolbox "Getting Started" manual [MathWorks, 2006]. Values for the rotor inertia, armature resistance, and so forth were used in a MATLAB model whose input was voltage and whose output was position. Even though these values are incorrect for the motor we will be using, they give us a starting point for design. The system was modeled in the Control System Toolbox, and the behavior was adjusted to suit the initial design requirements. The op-amps U3A and U3B as well as their network components were chosen based on this initial modeling session. They are simply placeholders to be changed as needed to get deeper into the design process.

2.5 Mechanical Design and Construction

The shaft repeater was initially constructed using an aluminum base plate with multiple "Tee" type slots along its length. This general type of extruded metal profile is common to many industries and allows components to be aligned easily. A perspective view of the base plate profile is shown in Figure 2.5. A 500-mm length was used for the prototype shown in Figure 2.3. Right-angle aluminum gussets were used to mount the motor and feedback potentiometer. These were machined as required to facilitate the component mounting. A swiveling tee nut system is used by the manufacturer to mount gussets to T-slots, and nuts having a ¼"-20 UNC tapped hole were used so that U.S. standard hardware could be used. A rubber spider coupling scheme was used to couple the shafts of the motor and potentiometer. Because the shafts were different sizes, 1/8-in. bore couplers were drilled out on a lathe to fit the shafts.

FIGURE 2.5
Perspective view of the base plate.

2.6 Parts List

Table 2.1 gives a complete parts list for the shaft repeater prototype system.

2.7 The Prototype as a Control System

As explained in Chapter 1, the DC servo is essentially a feedback control system whose input is a signal representing the commanded position of a load to be moved and whose output is a signal representing the actual position of that same load. As such, it can be represented in the terminology of control systems. The components of such a system are usually represented pictorially in block form, with lines connecting the blocks and variables assigned to their inputs and outputs. The block diagram system has become a universal way for engineers to communicate on a "system level," without necessarily going into all the details of a particular design.

Because the purpose of this book is to communicate details about DC servos, we will present the standard block diagram for a control system and use it to identify pieces of our shaft repeater design. This way we will intentionally avoid the generality of most systems-level work and get into the details of the DC servo design presented here in an orderly fashion. As stated earlier, the rationale for certain component selections will be the subject of later chapters.

TABLE 2.1

Shaft Repeater Parts List

Item	Designator	Description	Vendor/Part Number	Quantity
Electrical				
DC-DC converter	V2, V3	± 12 V @ 625 mA	Mean Well DKE15A-12	1
Op-amp IC	U1, U2, U3	TL082ACP	Jameco 214076	3
Transistor NPN	Q1	TIP41C	Jameco 179371	1
Transistor PNP	Q2	TIP42C	Jameco 139580	1
DC gear motor	M1	250 mA, 60:1 Gear ratio, 71 rpm	Jameco 155854	1
Potentiometer	POT1, POT2	1 K, single turn	Jameco 255628	2
Potentiometer	POT3	100 K, single turn	Jameco 255557	1
Capacitor	C1, C2	47 uF electrolytic	Jameco 11105	2
Capacitor	C3	2.2 uF Mylar film	Jameco 93999	1
Resistor	R11, R12, R13, R14, R22	10 KΩ, 1%, metal film	Mouser 271-10K-RC	5
Resistor	R18, R19	10 KΩ, 1%, metal film	Mouser 271-1K-RC	2
Resistor	R16	470 Ω, 1%, metal film	Mouser 271-470-RC	1
Resistor	R20	200 KΩ, 1%, metal film	Mouser 271-200K-RC	1
Resistor	R21	402 KΩ, 1%, metal film	Mouser 271-402K-RC	1
Mechanical				
Base plate	BASE1	22.5 × 180 profile	Bosch Rexroth 3 842 990 345/500mm	1
Motor mount	MOUNT1	60 × 60 gusset	Bosch Rexroth 3 842 523 546	1
Potentiometer mount	MOUNT2	60 × 60 gusset	Bosch Rexroth 3 842 523 546	1
Tee nut	NUT1	10 mm × ¼"-20 UNC tap	Bosch Rexroth 3 981 021 323	2
Motor shaft coupling	COUP1	.125 ID hub	Jameco 162288	1
Potentiometer shaft coupling	COUP2	.250 ID hub	Jameco 162288 (bored to .250-in. diameter)	1
Spider coupler	COUP3	Rubber spider	Jameco 162000	1

S1

Desired value ──V──⊗──E── G(D) ──C──→ Controlled variable
                    ⊟   Control
                          elements

Feedback

FIGURE 2.6
Control system block diagram—basic form.

2.8 Block Diagram Representations

Figure 2.6 is a block diagram of a control system in its most basic form. The block G represents the transfer function (made up of differential equations) relating the output C of the system's components (often called the "plant") to the error (or actuating) signal E. The error signal is the difference between the desired value V and the actual value C. The summing node S1 represents an addition or subtraction operation depending on the signs depicted. Similarly, blocks denote multiplication by the preceding variable. Thus, it should be readily apparent that

$$E = V - C \tag{2.1}$$

$$C = E \; x \; G(D) \tag{2.2}$$

The "x" operator for multiplication will be left out (and assumed) as appropriate throughout this book.

Figure 2.7 shows a generalized block diagram of a feedback control system using standard nomenclature. In this book we will continue to use this

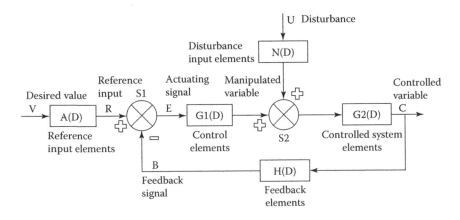

FIGURE 2.7
Control system block diagram—standard form.

nomenclature whenever possible. This standardization allows communication between engineers and non-engineers, as well as between engineers of different disciplines or concentrations. To more specifically guide our efforts here, Doebelin suggests the following:

> In studying a specific physical system...it is desirable to use the standard symbols V, R, C, etc., as subscripts on other symbols which indicate more clearly the physical nature of the variables involved. Thus, in a servo-mechanism, the reference input might be an angular rotation and would be designated as θ_R, while the controlled variable (also an angular rotation) would be called θ_C. This type of terminology has the advantage of retaining the essence of the standard notation, while also keeping the physical nature of the system apparent.

In Figure 2.7, we note the following standard subscripts and their meanings:

$$V \triangleq desired\ value$$

$$U \triangleq disturbance$$

$$R \triangleq reference\ input$$

$$E \triangleq actuating\ signal$$

$$M \triangleq manipulated\ variable$$

$$C \triangleq controlled\ variable$$

$$B \triangleq feedback\ signal$$

2.9 Electrical Schematic Walk-Through

We now have the necessary tools to "walk through" the electrical schematic of the shaft repeater given in Figure 2.4, identifying its constituent parts in terms of our control system block diagram, Figure 2.7, and doing some necessary circuit analysis along the way.

2.9.1 Reference Input Elements

The input to the circuit is the mechanical turning action of POT1 by the user and is represented by the desired value of shaft position, θ_V. The input circuit is shown in Figure 2.8. The reference input element is POT1, whose signal is conditioned by ballast resistors R18, R19, and op-amp U2B. The

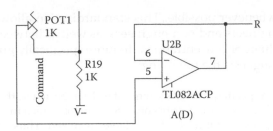

FIGURE 2.8
Reference input elements.

op-amp is connected as a buffer, so its gain is +1. The transfer function A(D) is computed as

$$A(D) = \frac{R}{V} \tag{2.3}$$

Examining the potentiometer bridge, involving the parallel combination of POT1 and POT2, the voltage across both pots is equal. Each wiper is used as a pick-off point to translate their rotary motions into voltages. The voltage across each pot is

$$V_P = V_{TOTAL}\left(\frac{1K\|1K}{[1K+1K+(1K\|1K)]}\right) \tag{2.4}$$

V_{TOTAL} is the potential difference between the positive and negative power supplies (see Figure 2.13), so in our case

$$V_{TOTAL} = +12V - (-12V) = 24V \tag{2.5}$$

and

$$V_P = 24V\left(\frac{500}{2500}\right) = 4.8V \tag{2.6}$$

The potentiometers are specified for a 300° rotational span, and we will express their transfer functions in units of volts/radian so that

$$A(D) = \frac{R}{V} = 4.8V\Big/\left[300°\left(2\pi\frac{rad}{360°}\right)\right] \tag{2.7}$$

$$A(D) = 0.92\frac{V}{rad} \tag{2.7a}$$

We note that $A(D)$ does not involve a differential equation, so it is commonly referred to as a "zero-order" system.

2.9.2 Summing Junction

The "summing junction" is a loosely used term in block diagrams to indicate up to four combinations of addition and subtraction of signals. The standard feedback control system of Figure 2.7 shows two junctions, S1 and S2. In the case of our shaft repeater, S2 is not used, simply because we choose not to include a disturbance input in our model. We do not expect a disturbance under the circumstances included in this book. Of course, if the reader intentionally impedes motion (and this is not advised for safety reasons), modeling that disturbance would be justified.

S1 is a difference junction and the circuit used to realize it is shown in Figure 2.9. Op-amp U1B is configured as a difference amplifier with all resistors of equal value. Its output is the actuating signal *E*. The buffers previously mentioned are used in this circuit to prevent "loading" of the input circuit by the summing junction. For more information on op-amp circuits including the loading problem, [Franco, 1988] and [Mancini, 2003] are both excellent references.

2.9.3 Control Elements

We have two stages of control elements, realized by op-amps U3A and U3B, and shown in Figure 2.10. The first stage is a compensation network, and the second stage is an intermediate gain stage. Together these represent the block G1(D) of Figure 2.7. Let us examine these in more detail.

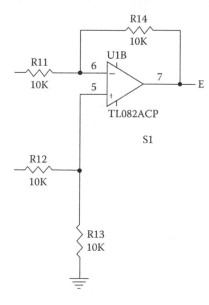

FIGURE 2.9
Summing junction elements.

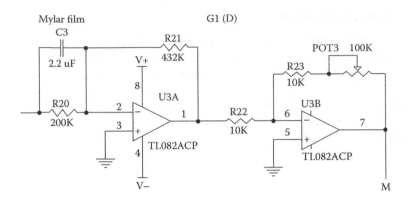

FIGURE 2.10
Control elements.

Op-amp U3A is configured as a "lead network," and the reason for this choice will become apparent later. The circuit is a differentiator, taking the approximate derivative of the input signal E. Differentiators are called lead networks because the phase angle of the output leads that of the input if the input is sinusoidal. An important point here is that phase angle is arbitrary and has nothing to do with conservation of energy. Therefore, having an output leading an input in terms of phase angle is reasonable in nature although it may seem counterintuitive.

The lead network transfer function can be shown to be [Doebelin, 1962]

$$\frac{V_O}{V_I}(D) = -\frac{R_{21}}{R_{20}}(R_{20}C_3D + 1) \tag{2.8}$$

This function is first order. The intermediate gain stage has a DC gain of $-(R_{23} + POT3)/R_{22}$ and is another zero-order function. The gain can be varied over roughly a factor of 10 for future experimentation. Combining these, we have for the control elements

$$G1(D) = \frac{M}{E} = \frac{(R_{23} + POT3)}{R_{22}} \frac{R_{21}}{R_{20}}(R_{20}C_3D + 1) \tag{2.9}$$

2.9.4 Disturbance and Disturbance Input Elements

As mentioned earlier, we do not expect the shaft repeater to be disturbed in operation. For the purposes of this example, it will be used with no mechanical load except the feedback potentiometer POT2. The environment is academic and thus is fairly benign. For now, we can eliminate U, $N(D)$, and S2 in Figure 2.7 from consideration.

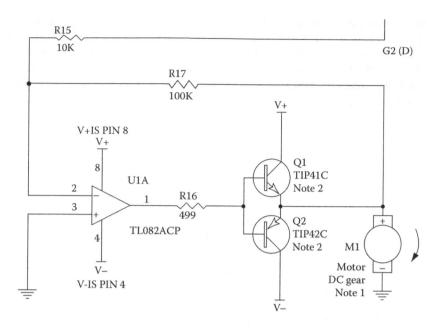

FIGURE 2.11
Controlled system elements.

2.9.5 Controlled System Elements

We make a somewhat arbitrary choice of our power amplifier and motor as the controlled system elements, as shown in Figure 2.11. This is done primarily because the currents in this part of the circuit are boosted by roughly a factor of 100. In general, we expect the motor M1 to draw in excess of 100 mA during motion.

The op-amp U1A is connected as an inverting amplifier of gain −10, with a current booster stage inside its feedback loop. The current booster is configured as a class B push-pull amplifier. As presented, its output will be distorted (commonly known as "crossover distortion"). [Widlar, 1968] advises that "although this circuit does have a dead zone [distortion], it can be neglected at frequencies below 100 Hz because of the high gain of the amplifier." The bandwidth of the mechanical side of the shaft repeater is typically in the 10 Hz range. The resistor R16 is not strictly necessary but is usually used to prevent high-frequency parasitic capacitive oscillations of the output transistors. Oscillations such as these will be filtered out by the mechanics of the system. Widlar also advises that "adequate bypassing should be used on the collectors of the output transistors to ensure that the output signal is not coupled back into the amplifier." This is reasonable since the higher currents in the motor leads will produce magnetic fields capable of inducing currents in nearby wires, especially the amplifier input leads which are a high-impedance area.

(A high-impedance area just takes a small current to affect its signal.) C1 and C2, placed close to Q1 and Q2, are used to accomplish the decoupling function and are sized according to the allowable currents and the fastest signal capable of affecting the servo. For example, if we say that the motor draws 250 mA at maximum efficiency under load, and the fastest command input we expect is 5 V/ms, then

$$I = C \frac{dV}{dt} \tag{2.10}$$

and

$$C = I / \left(\frac{dV}{dt} \right) \tag{2.11}$$

So the capacitors end up being approximately 0.25 A/(5 V/.001 sec) = 50 μF, and should be rated for 3 or 4 times the supply voltage, say 50 V. In this size and voltage range, a polarized type electrolytic capacitor is most easily obtained. Current limiting is provided by the DC-to-DC converter used in the power supply, so the 50 Ω resistors in series with each collector suggested by are not necessary, and would make the output stage less efficient.

Referring once again to the system block diagram, Figure 2.7, the power amplifier just described is much faster than the mechanics of the system. As before, we model it as a constant gain term with transfer function

$$\frac{e_{amp}}{M}(D) = K_{amp} = -10 \tag{2.12}$$

The transfer function for the geared motor M1 is first order and can be expressed as

$$\frac{\omega_C}{e_{amp}}(D) = \frac{K_{gear}}{\tau_{gear}(D) + 1} = \frac{output}{input} \tag{2.13}$$

This function will be derived in detail later. Because the controlled variable in our shaft repeater is position (not velocity), the final element to complete the G2(D) block is a change from velocity to position. At first glance this change may seem straightforward, but there are subtleties here that deserve explanation. The output shaft velocity is the derivative of its position, so that

$$\omega_C = \frac{d\theta c}{dt} = D\theta_C \tag{2.14}$$

This means that the output shaft position is the integral of its velocity. Mathematically, the definition of integration using the D notation is

$$\int_o^t \omega_C dt \triangleq \left(\frac{1}{D} \right) \omega_C \tag{2.15}$$

Using this definition, dividing both sides of Equation (2.14) by the *D*-operator gives the velocity integral

$$\left(\frac{1}{D}\right)\omega_C = \theta_C \tag{2.16}$$

Dividing both sides of Equation (2.16) by ω_C, the transfer function between position and velocity is

$$\frac{\theta_C}{\omega_C} = \frac{1}{D} \tag{2.17}$$

So when control system engineers say "the motor is an integrator," they mean that if a constant voltage is applied to a DC motor, its output position will increase linearly with time; that is, it will behave as the output of an integrator. Finally, we can combine Equations (2.12), (2.13), and (2.17) to get the overall transfer function G2(*D*), which is

$$G2(D) = \frac{\theta_C}{M} = \frac{K_{amp}K_{gear}}{(\tau_{gear}D+1)D} \tag{2.18}$$

2.9.6 Feedback Elements

The circuit shown in Figure 2.12 has the exact same transfer function as the reference input elements, so that from (2.7a) we have

$$H(D) = \frac{B}{C} = 0.92\frac{V}{rad} \tag{2.19}$$

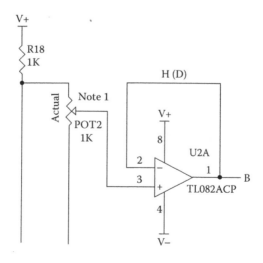

FIGURE 2.12
Feedback elements.

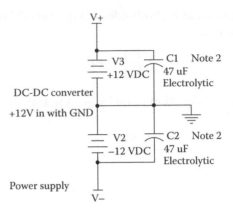

FIGURE 2.13
Power supply elements.

2.9.7 Power Supply Elements

A DC-to-DC converter was used in this system to convert a single +12V supply to split supplies suitable for the op-amp and power amplifier operation. Its cost is justified by the simplicity gained in our design effort. These converters are highly specialized switching power supplies [Pressman, 1991] whose design is well beyond the scope of this book. Modules such as the one used here are so commonly used in industry today that it is much easier to buy the technology rather than reinvent it. Figure 2.13 shows the outputs of the converter schematically as batteries, with the bypass capacitors to be placed near Q1 and Q2.

References

Doebelin, E., *Dynamic Analysis and Feedback Control*, McGraw-Hill, 1962.

Foster, C., *Real Time Programming—Neglected Topics*, Addison-Wesley, 1981.

Franco, S., *Design with Operational Amplifiers and Analog Integrated Circuits*, McGraw-Hill, 1988.

Mancini, R., *Op Amps for Everyone*, Newnes, 2003.

MathWorks, The, "Getting Started with the Control System Toolbox," 2006.

Pressman, A., *Switching Power Supply Design*, McGraw-Hill, 1991.

Widlar, R., "Monolithic Op Amp—The Universal Linear Component," National Semiconductor AN-4, 1968.

3

DC Motors in Servo Systems

3.1 Introduction

The vast majority of DC motors used in servo-controlled systems today are fractional-horsepower, permanent-magnet types. Three dominant factors have influenced this: (1) the development over the past 50 years or so of permanent magnet materials having both high residual flux and high coercive force; (2) the commercial market developed over the last 30 years for computer peripheral devices, requiring motion of one kind or another, largely driven by the revolution in portable computing power and the explosive growth in information storage and graphics reproduction; and (3) the development of power electronics capable of driving these motors and controlling them efficiently. The availability of this "motion control" technology, pioneered roughly between 1970 and 1985, has in turn driven small DC motors into many other markets, including consumer electronics and automotive accessories. Table 3.1 is an attempt to summarize these trends as they exist today.

3.2 Operational Principles

A simple rotating DC motor is shown in Figure 3.1. It consists of a multiturn rectangular wire coil rotating in space around a fixed axis at an angular speed ω. The rotating part of the machine is the rotor, whereas the stationary part is the stator. A magnetic flux density B is supplied by north and south poles located on the stator. In order to minimize the reluctance in the motor's magnetic circuit, the rotor windings are usually interspersed between slots in a stack of iron laminations. Also to minimize reluctance, as small an air gap as possible is maintained between the rotor and stator of the machine. To keep the magnetic flux in the air gap constant over the area of the pole faces, these faces are curved. If rotated by another motor, the machine of

TABLE 3.1

The Utility of DC Servos Today

Societal Need	Some Specific Applications	Some Organizations Involved
Aircraft transport	Instrument servos, stability augmentation, auto-pilot systems	Boeing, McDonnell-Douglas, Airbus
Automotive transport	Dashboard gauges, seat positioning, convertible top control, automatic entry, automatic steering	Daimler-Benz, Ford, GM, Volvo, Toyota, Honda, Chrysler, Nissan, Mitsubishi, Hyundai
Consumer electronics	CD/DVD (3 servos/unit), still cameras (2 servos/unit), video cameras (3 servos/unit)	Sony, Philips, Matsushita, Olympus, Kodak, Nikon, Canon, Konica-Minolta, Bose
Defense systems	Weapon aiming, radar tracking, missile guidance, unmanned vehicles, robotic engagement	General Dynamics, Loral, Raytheon, Ball Aerospace, Jet Propulsion Lab (CalTech), Lincoln Lab (MIT)
Health care	Patient positioning tables, kidney dialysis machines, infusion pumps, chart recorders, robot-assisted surgery, home health care	GE, Siemens, Haemonetics, Baxter Healthcare, Agilent Technologies
Information storage	CD-R/DVD-R (3 servos/unit), hard disk drives (2 servos/unit)	Sony, Philips, Matsushita, Seagate, Connor, HP, EMC, IBM
Manufacturing	Machine tools, industrial robots, plastic molding machines	Mori Seiki, Rong Fu, Epson, Fanuc, Denso
Printed media	Copiers, ink-jet printers, laser printers	HP, Canon, Xerox, Kodak, Brother, Epson
Space exploration	Life-support systems, navigation systems, remote soil sampling, remote vehicle positioning, scanning spectrometers, terrestrial telescope drives	General Dynamics, Loral, Raytheon, Ball Aerospace, Jet Propulsion Lab (CalTech), Lincoln Lab (MIT), Meade Instruments, Celestron
Education and toys	Radio-controlled aircraft, small robots	Hitec, Futaba, Lego, Parallax

Figure 3.1 would generate an open circuit AC output if it were not for the two brushes and two commutator segments. This "split ring commutator" arrangement of rotating switches allows the current in the coil to be reversed when the magnetic flux changes direction relative to the plane of the coil as it rotates. In this manner, an open circuit DC voltage e_i can be produced from an applied unidirectional torque $T_M(t)$. If the circuit is closed, and a DC voltage e_i applied, this process is reversible (with resistance losses), so that a unidirectional torque $T_M(t)$ can be produced by the machine. This connection switching process is known as commutation and is the basis of all DC motor

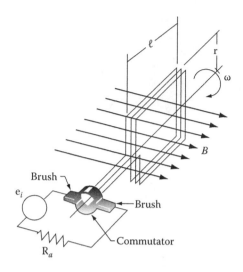

FIGURE 3.1
Schematic for a simple DC motor with split ring commutator.

operation. A small DC motor, a few inches long, is shown disassembled in Figure 3.2. For smooth operation, the commutator is a multisegment assembly known as the armature. In this case, there are 13 armature segments, 13 rotor slots, and two brushes 90° apart.

There are two basic natural phenomena responsible for explaining the action of DC motors. One is the Lorentz force and the other is Faraday's law of induction. The Lorentz equation predicts the force exerted on an electric charge moving in a magnetic field. For a single coil of wire in Figure 3.1, it can be shown that the torque exerted on the coil is

$$T_M(t) = 2ri_a(t)lB \tag{3.1}$$

Faraday's law predicts the voltage induced in an electric circuit moving in a magnetic field. For the same coil of wire, it can be shown that the voltage induced in the coil is

$$e_b(t) = \left(\frac{2}{\pi}\right) A_p B \omega(t) \tag{3.2}$$

where the subscript b indicates the so-called back emf generated (because it tends to oppose the current that caused it) and A_p is the area under each pole face. These two relations (3.1) and (3.2) give us necessary tools for further analysis of the different DC motor types.

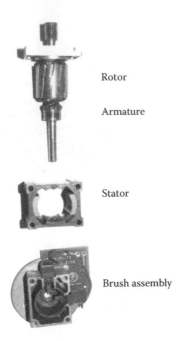

Rotor

Armature

Stator

Brush assembly

FIGURE 3.2
Major components of a small brushed PMDC motor. Note skewing of rotor slots, multiple armature segments, and curvature of stator magnets. The number of rotor slots (13 in this particular motor) equals the number of armature segments.

3.3 Basic Classes of DC Motors

Most DC motors used in servo applications, especially positioning servos, are fractional horsepower (<1 HP) machines. Although there are many speed control applications for DC machines above 1 HP, they are becoming less popular due to the availability of variable-frequency AC drives. For the purposes of this book, we will concentrate on DC motors of <1 HP which generally weigh less than 25 lb. Figure 3.3 gives a further breakdown of DC motor classes, which will now be briefly discussed.

3.3.1 Brushed vs. Brushless Motors

The majority of DC position servos today are implemented with brushed motors of the basic type depicted in Figure 3.2. Velocity servos, especially at high speeds, are increasingly being designed using the somewhat newer "brushless DC" (BLDC) motor variant. These motors are inherently different from most DC motor varieties, more closely resembling step motors. Because of the many differences in construction and drive schemes, BLDC motors are discussed in detail elsewhere [Chapman, 1999; Stiffler, 1992].

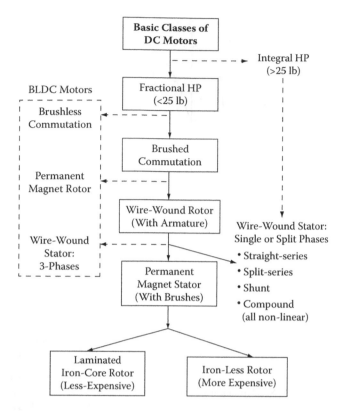

FIGURE 3.3
Basic classes of DC motors.

3.3.2 Wound Field Motors

The magnetic field flux density B shown in Figure 3.1 can be generated using either wound coils (electromagnets) or permanent magnets. There are a few different variations of field windings, especially in the integral horsepower range. These include straight-series, split-series, shunt, and compound windings. An excellent review of these is given by Chapman. The shunt motor, with a separate power supply used for the field winding (preferably a current source), is functionally equivalent to the permanent magnet motor.

3.3.3 Permanent Magnet Motors

Advances in magnetic material development over the past 30 years have revolutionized the permanent magnet DC (PMDC) motor and revitalized its use in small servo applications. The invention of the so-called rare earth materials, containing the elements samarium and neodymium, has made this revolution possible. Figure 3.4(a) shows a magnetization curve for a

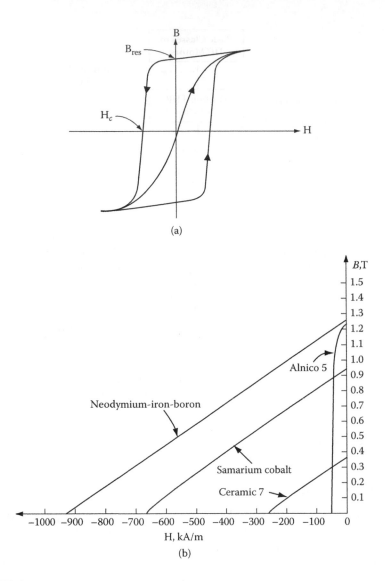

FIGURE 3.4
(a) Generic magnetization curve for a magnetic material; (b) second quadrant of magnetization curve for four permanent magnet materials.

typical magnetic material. When a strong magneto-motive force (MMF) is applied to the material and then removed, a residual flux B_{res} will remain. To force this residual flux to zero, an MMF of opposite polarity, greater than or equal to the coercive magnetizing intensity H_C, needs to be applied. This process establishes the second quadrant of the *B-H* curve. Chapman notes that for normal machine applications, such as the rotor laminations of a DC motor, a material is chosen to have a small B_{res} and H_C to minimize magnetic hysteresis losses. Conversely, a material for the poles of a PMDC motor must ideally have both a large B_{res} and a large H_C. This provides a high flux density in the air gap and good resistance to demagnetization by the rotor MMF when the rotor current is highest—for instance, in a motor stall or start-up condition. Figure 3.4(b) shows the second-quadrant performance of traditional, older materials as well as the newer rare earth materials. Stiffler confirms that Alnico 5 is used in relatively low-torque applications, whereas samarium-cobalt is the material of choice in most high-performance PMDC motors, since it will not demagnetize under high-current conditions.

3.3.4 The Fractional Horsepower Brushed PMDC Motor

For position servos using motion profiling for point-to-point moves, the fractional-horsepower brushed PMDC motor remains the first choice in industry today, although the BLDC motor is becoming less expensive to manufacture and drive. The main advantages of PMDC motors in control applications are as follows:

1. Linear torque-speed characteristic
2. High stall and accelerating torque
3. Magnetic flux generated without need for electric power
4. Smaller frame size and lighter weight for a given power output

Two basic classes of brushed PMDC motors exist to satisfy the range of possible servo applications, and the differences relate to rotor construction and efforts to increase the motor's torque-to-inertia ratio. The laminated iron-core rotor previously discussed (Figure 3.2) is by far the most widely used [Kuo, 1982] in servo applications with moderate performance demands, since the rotor structure has a high thermal capacity and can take overloads for extended periods of time without damage. Its disadvantages include high moment of inertia, high inductance, and "detent" positions due to unequal reluctance paths. This tendency of the rotor to assume preferred positions, which leads to torque ripple in operation, is reduced by skewing the rotor slots. Motors of the iron-less rotor class include the disk or "pancake" type and the cup or "basket" type. Both have higher torque-to-inertia ratios than iron-core motors, but both are also much more expensive.

3.4 Considerations in Motor Selection

Motor selection is a process involving many variables, including the characteristics of the load to be driven. To keep our discussion focused, we will limit our motor choices to brushed PMDC motors used in the fractional

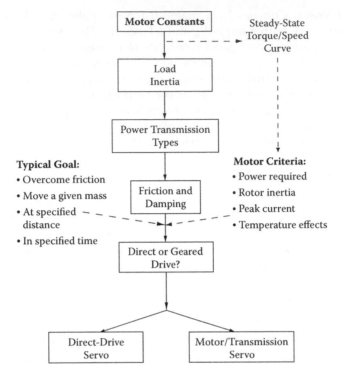

FIGURE 3.5
PMDC servo motor selection criteria.

horsepower range. Figure 3.5 is an attempt to organize the thought process involved in choosing a motor for a particular application, and we will now look into the details of this selection process.

3.4.1 Motor Constants

For a constant value of flux density under the pole faces of a PMDC motor of a particular size, Equations (3.1) and (3.2) reduce to

$$T_M(t) = K_T i_a(t) \tag{3.3}$$

$$e_b(t) = K_e \omega(t) \tag{3.4}$$

where K_T is the torque constant of the motor and K_e is the back emf constant.

3.4.2 Steady-State Torque/Speed Curve

The torque/speed curve is a fundamental performance metric for electric motors [Electro-Craft, 1980]. For the PMDC motor, the relationship between torque and speed is linear for a given input voltage. Recalling the dynamic model introduced in Chapter 1, for the electrical circuit we again have

$$i_a R_a + K_e \omega_o = e_i \tag{3.5}$$

For Newton's law at the motor shaft, we now assume a constant friction load independent of velocity is applied that is much larger than the viscous friction assumed in Chapter 1. The viscous (damping) friction $B\omega_o$ being negligible gives

$$K_T i_a - T_F = J \dot{\omega}_o \tag{3.6}$$

In the steady state, the motor is running at constant velocity and the acceleration is zero. From (3.6) we now get

$$i_a = T_F / K_T \tag{3.7}$$

Substituting (3.7) into (3.5) we have the linear relation

$$\frac{T_F R_a}{K_T} + K_e \omega_o = e_i \tag{3.8}$$

From (3.8), we get the no-load speed

$$\omega_{NL} = e_i / K_e \tag{3.9}$$

and stall torque

$$T_{FS} = e_i K_T / R_a \tag{3.10}$$

A family of torque/speed curves can be drawn for a given motor as shown in Figure 3.6, and such curves are usually supplied by motor manufacturers.

3.4.3 Rotor Inertia

The moment of inertia of a rigid body undergoing pure rotational motion about a fixed axis is defined from Newton's law in the rotational form

$$\sum Torques = J\dot{\omega} \tag{3.11}$$

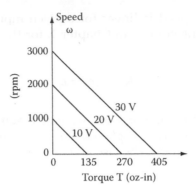

FIGURE 3.6
Torque/speed curves for three terminal voltages.

where J is the moment of inertia. The rotor of a PMDC motor is a right circular cylinder as shown in Figure 3.7. Although most rotors are not completely homogeneous, this is a reasonable assumption for estimating the inertia of a rotor when none is given. Following [Doebelin, 1972], we can apply the translational form of Newton's law to a ring-shaped mass element of infinitesimal width dr at radius r. Since every particle in this mass has the same tangential acceleration,

$$Tangential\ force = (2\pi rL\rho dr)(r\dot{\omega}) \tag{3.12}$$

The torque on the element is r times the force, so the total torque on the cylinder is

$$T = \int_0^R 2\pi\rho L\dot{\omega}r^3 dr = \left(\frac{MR^2}{2}\right)\dot{\omega} \tag{3.13}$$

From (3.11) we finally get for the right circular cylinder

$$J = \frac{MR^2}{2}, N\cdot m\cdot sec^2 \tag{3.14}$$

The moment of inertia for other shapes is calculated in a similar fashion and can be found in Doebelin, as well as most rigid body dynamics texts.

3.4.4 Power Transmission to a Given Load

DC servo systems almost always involve some type of power transmission means, since PMDC motors are inherently high-speed, low-torque devices,

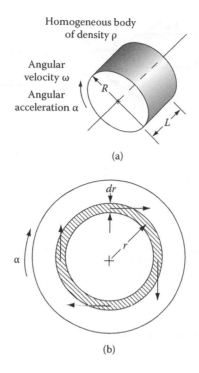

(a)

(b)

FIGURE 3.7
Diagram used in rotational inertia calculation for a right circular cylinder; (a) cylinder as a whole; (b) infinitesimal cylindrical segment of the calculus.

whereas the load to be driven in a typical application requires high torque to achieve rapid acceleration. We now proceed to examine three such transmissions: gear train, belt-pulley, and lead screw drives.

3.4.4.1 Gear Train Drive

A typical gear mesh reduction for a DC servomechanism is shown in Figure 3.8(a), where the motor shaft is subscripted M and the output or load shaft is subscripted L. This is a very common arrangement, where the motor is driving a purely inertial load with no added load torque applied. Assuming that the gears are rigid bodies and have no backlash and that the shafts are rigid, the two rotations are related by a gear "reduction" ratio such that

$$\theta_M = n\theta_L \tag{3.15}$$

where n is by definition greater than unity. For example, if $n = 10$, the motor shaft would turn 10 revolutions for every revolution of the load shaft. We further assume the presence of viscous damping factors B_M and B_L due to

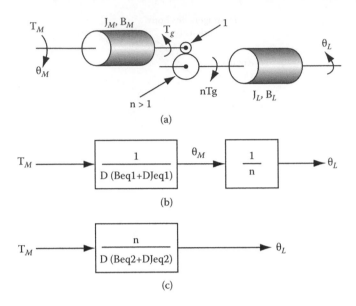

FIGURE 3.8
(a) Gear-reduced system; (b) lumped parameters referred back to motor; (c) lumped parameters referred forward to load.

bearing and gear lubrication. An analysis of the free body diagrams of both shafts results in the following differential equations:

$$T_M - T_g - B_M\dot{\theta}_M = J_M\ddot{\theta}_M \tag{3.16}$$

$$nT_g - B_L\dot{\theta}_L = J_L\ddot{\theta}_L \tag{3.17}$$

There are two fundamental methods [Brown and Campbell, 1938] for simultaneous solution of the preceding two equations. One is to refer or "reflect" the inertia and damping of the load back to the motion of the motor shaft, which is useful in motor selection for a given application. The other is to refer the inertia and damping of the motor to the motion of the load shaft. This is useful in the creation of the overall system transfer function, where predicting the output motion is of primary importance. As we shall see, the two methods result in the same transfer functions and are interchangeable with a minimum of effort. The only difference lies in the reader's perspective and the task at hand. To simplify the analysis, we can invoke the D-operator in the preceding equations to get

$$T_M - T_g - B_M D\theta_M = J_M D^2\theta_M \tag{3.18}$$

$$nT_g - B_L D\theta_L = J_L D^2\theta_L \tag{3.19}$$

Method 1 eliminates the torque shared by the two gears T_g and reflects all elements back to the motor shaft. Using (3.15) and (3.19) to solve for T_g in terms of θ_M we get

$$T_g = \frac{D(B_L + J_L D)}{n^2} \tag{3.20}$$

Substituting this expression into (3.18) and collecting terms,

$$T_M = D\theta_M[(B_M + B_L/n^2) + D(J_M + J_L/n^2)] \tag{3.21}$$

If we define input T_M and output θ_M, the transfer function for the motor and load in terms of the motor shaft motion is

$$\frac{\theta_M}{T_M}(D) = 1/D(B_{eq1} + DJ_{eq1}) \tag{3.22}$$

With subscript "1" referring to our first method of solution, we have the lumped parameters

$$\textit{equivalent damping} \triangleq B_{eq1} = B_M + B_L/n^2 \tag{3.23a}$$

$$\textit{equivalent inertia} \triangleq J_{eq1} = J_M + J_L/n^2 \tag{3.23b}$$

An equivalent block diagram for this case is shown in Figure 3.8(b), where a second block containing the gear ratio is placed in cascade with the first block to transfer the lumped parameters (3.23) to the system output motion θ_L. Converting (3.21) and (3.23) to the time domain,

$$T_M - B_{eq1}\dot{\theta}_M = J_{eq1}\ddot{\theta}_M \tag{3.24}$$

This equation is of the familiar form

$$\textit{driving torque} + \textit{damping torque} = \textit{inertia} \times \textit{acceleration} \tag{3.24a}$$

The transfer function (3.22) represents a first-order system with a pure integration in cascade, of the form

$$KG(D) = \frac{K_1}{(D(\tau_1 D + 1))} \tag{3.25}$$

where

$$K_1 = 1/B_{eq1} \qquad (3.26a)$$

$$\tau_1 = J_{eq1}/B_{eq1} \qquad (3.26b)$$

Method 2 eliminates the torque shared by the two gears T_g and refers all elements forward to the load shaft. Using (3.15) and (3.18) to solve for T_g in terms of θ_L we get

$$T_g = T_M - D(B_M + J_M D)n\theta_L \qquad (3.27)$$

Substituting this expression into (3.19) and collecting terms,

$$nT_M = D\theta_L[(n^2 B_M + B_L) + D(n^2 J_M + J_L)] \qquad (3.28)$$

If we define input T_M and output θ_L, the transfer function for the motor and load in terms of the load shaft motion is

$$\frac{\theta_L}{T_M}(D) = n/D(B_{eq2} + DJ_{eq2}) \qquad (3.29)$$

With subscript "2" referring to our second method of solution, we have the lumped parameters

$$\text{equivalent damping} \triangleq B_{eq2} = n^2 B_M + B_L \qquad (3.30a)$$

$$\text{equivalent inertia} \triangleq J_{eq2} = n^2 J_M + J_L \qquad (3.30b)$$

An equivalent block diagram for this case is shown in Figure 3.8(c), where the lumped parameters (3.23) are directly transferred to the system output motion θ_L. Converting (3.28) and (3.30) to the time domain,

$$nT_M - B_{eq2}\dot{\theta}_L = J_{eq2}\ddot{\theta}_L \qquad (3.31)$$

This equation is also of the familiar form

$$\text{driving torque} + \text{damping torque} = \text{inertia} \times \text{acceleration} \qquad (3.31a)$$

The transfer function (3.29) represents a first-order system with a pure integration in cascade, of the form

$$KG(D) = \frac{K_2}{(D(\tau_2 D + 1))} \qquad (3.32)$$

where

$$K_2 = n/B_{eq2} \tag{3.33a}$$

$$\tau_2 = J_{eq2}/B_{eq2} \tag{3.33b}$$

It is natural that the time constants are equal,

$$\tau_1 = \tau_2 \tag{3.34}$$

since the two models we have created above are equivalent and physically identical.

3.4.4.2 Belt-Pulley Drive

The belt and pulley system is one basic method of translating the rotational motion of a PMDC motor to linear motion, a popular example being the optical scanning mechanism of a flatbed scanner or photocopier. This arrangement is shown schematically in Figure 3.9. Applying Newton's law to the mass M,

$$force = \frac{T}{r} = M\ddot{x} = M(r\dot{\omega}) \tag{3.35}$$

The torque applied at the pulley is r times the force applied to the mass, so that

$$torque = rM(r\dot{\omega}) = Mr^2\dot{\omega} \tag{3.36}$$

The equivalent inertia of this system is simply the sum of the motor and load inertias, so that

$$J_{eq} = J_M + Mr^2 \tag{3.37}$$

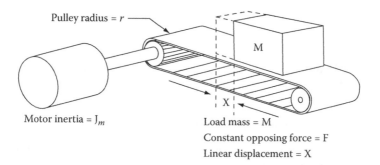

FIGURE 3.9
Belt-pulley drive system.

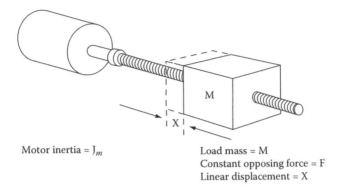

Motor inertia = J_m

Load mass = M
Constant opposing force = F
Linear displacement = X

FIGURE 3.10
Lead screw drive system.

3.4.4.3 Lead Screw Drive

Smaller machine tools use PMDC motors as actuators, and the lead screw mechanism is the mainstay of most machine tool movement—for example, the horizontal slide of a milling machine. The arrangement is shown schematically in Figure 3.10. The screw is characterized by its lead, which equals the length it is advanced by a single rotation. Relating the angular velocity of the screw to the linear velocity of the mass, we have

$$\omega = \left(\frac{2\pi}{L}\right)\dot{x} \tag{3.38}$$

Since $\dot{x} = r\omega$, the lead screw is equivalent to a belt-pulley drive of radius

$$r_{eq} = L/2\pi \tag{3.39}$$

For the lead screw drive,

$$J_{eq} = J_M + M\left(\frac{L}{2\pi}\right)^2 \tag{3.40}$$

A summary of these transmissions and their key parameters is given in Table 3.2. This will serve as a motor selection aid, since all parameters are referred back to the motor shaft.

3.4.5 Mechanical Friction and Damping

Fundamentally, friction is the resistance of an object to move when in contact with another object. In mechanics, the force of friction is described as

$$F_f = \mu N \tag{3.41}$$

TABLE 3.2

Key Parameters of DC Servo Transmissions, Referred Back to Motor Shaft

Transmission Type	Gear Drive	Pulley-Belt Drive	Lead Screw Drive
Load step size	θ_o	x	x
Coupling parameter	Gear ratio n	Pulley radius r	Screw lead L Define $r = L/2\pi$
Load	Inertial load J_L Opposing torque T_L	Mass M Opposing force F	Mass M Opposing force F
Rotation referred to motor shaft	$\theta_i = n\theta_o$	$\theta_i = x/r$	$\theta_i = x/r$
Torque referred to motor shaft	$T = T_L/n$	$T = Fr$	$T = Fr$
Inertia referred to motor shaft	$J = J_L/n^2$	$J = Mr^2$	$J = Mr^2$
Inertial match for max. load acceleration	$n^2 = J_L/J_m$	$r^2 = J_M/M$	$L^2 = (2\pi)^2 J_M/M$

where N is the force normal to the surfaces in contact and μ is the coefficient of friction. Frictional forces can further be divided into two basic categories, sliding friction and viscous friction.

3.4.5.1 Sliding Friction

There are two components involved in sliding friction. The first is static friction (or "stiction"), a retarding force that tends to prevent motion from beginning. This force exists only when velocity is zero and thus is completely nonlinear. The force-velocity relation is shown in Figure 3.11(a). The magnitude of static friction is always greater than its closely related counterpart, Coulomb friction, whose force-velocity relation is shown in Figure 3.11(b). Coulomb friction is constant with velocity and changes sign depending on

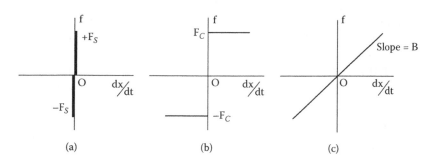

(a) (b) (c)

FIGURE 3.11
(a) Static friction; (b) Coulomb friction; (c) viscous friction.

the direction of velocity. It is also nonlinear and can be described mathematically as

$$f(t) = F_c \left[\frac{\frac{dx}{dt}}{\left| \frac{dx}{dt} \right|} \right]$$ (3.42)

3.4.5.2 Viscous Friction

Viscous friction (or "damping") is a retarding force that increases linearly with velocity. Its force-velocity relation is shown in Figure 3.11(c) and is described mathematically as

$$f(t) = B \frac{dx}{dt}$$ (3.43)

Viscous friction is usually caused by lubricants used between parts of a moving assembly, or windage, where a rotating part might experience resistance to movement due to the air surrounding it being perturbed in some predictable way.

3.5 Procedure for Meeting a Design Goal

In a typical application, a motor must be selected to overcome friction and move a mass through a certain distance in a specified time. The mass to be moved includes the mass of the motor's rotor, which cannot be ignored in geared-down systems [Doebelin, 1962]. From (3.23) we know that with step-down gearing, *equivalent inertia* $\triangleq J_{eq} = J_M + J_L/n^2$. For instance, if motor inertia $J_1 = 1$, load inertia $J_2 = 100$, and gear ratio $n = 100$, the load inertia appears to be 100 times larger than the motor inertia. However, because of the high (but not unusual) gear ratio, the inertial effect of the load is $1/(100)^2 = 1/10,000$, which is 100 times smaller than the motor. The lesson learned here is that motor inertia may be the dominant inertia in a servo system, and designers often will choose motors with the lowest rotor inertia possible.

Now suppose that we have a known load mass to be moved (or load inertia to be rotated) in a known time. A velocity profile is then assumed in order to proceed with motor selection calculations. Velocity profiling is a controllable parameter in a point-to-point move. The following numerical example might represent the line-by-line paper advance function of a printer, driven by a belt and pulley mechanism:

Print speed: 1000 lines per minute

Pulley radius: $r = 0.0127$ m

Load inertia: $J_L = 3.5 \times 10^{-5}$ N-m-sec²

Motor inertia: $J_M = 3.5 \times 10^{-6}$ N-m-sec²

Paper displacement: $d = 0.0032$ m

Start-stop time: $t_{ss} = 10$ milliseconds

Dwell time: $t_d = 20$ milliseconds

Load friction torque: $T_F = 0.07$ N-m

For simplicity, we assume the velocity profile to be triangular, consisting of two ramps as shown in Figure 3.12. Other profiles are possible, the most popular being a trapezoidal shape because it is more efficient in longer moves. Both profiles can be implemented in the command structure of a digital microcontroller. For the interested reader [Kuo and Tal, 1978] gives a very detailed treatment of how the choice of a velocity profile affects overall system efficiency. For our simpler example, it takes half the start-stop time to accelerate the load and half the start-stop time to decelerate it. The angular displacement is

$$\theta = \frac{d}{r} = \frac{0.0032}{0.0127} = 0.25 \; radians \tag{3.44}$$

The average angular velocity over the start-stop motion is

$$\omega_{avg} = \frac{\theta}{t_{ss}} = \frac{0.25}{0.01} = 25 \; rad/sec \tag{3.45}$$

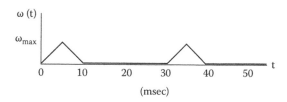

FIGURE 3.12
Velocity profile for the paper feed mechanism of a printer.

From Figure 3.12, the peak angular velocity is just twice the average, so that

$$\omega_{pk} = 50 \; rad/sec \qquad (3.46)$$

The angular acceleration is constant for a linearly increasing velocity and takes place over the first half of each move:

$$\dot{\omega} = \frac{\omega_{pk}}{\frac{t_{ss}}{2}} = \frac{50}{0.005} = 10,000 \; rad/sec^2 \qquad (3.47)$$

The torque needed to be overcome at the load by the motor is the sum of the torque needed to accelerate the load, the motor's rotor, and the load friction torque:

$$T_L = (J_M + J_L)\dot{\omega} + T_F = 3.85 \times 10^{-5}(10,000) + .07 = 0.36 \; N \cdot m \qquad (3.48)$$

Now that we know the torque and speed requirements at the load, we can proceed to select a motor for this application. The peak power needed at the load is

$$P_{L(pk)} = T_L \omega_{pk} = 0.36(50) = 23 \; N \cdot m/sec \qquad (3.49)$$

The torque/speed curve of a typical PMDC motor is shown in Figure 3.13. The power/speed curve is also shown, which for a linear torque/speed

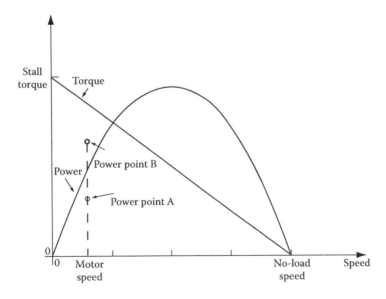

FIGURE 3.13
Performance characteristics of a brushed PMDC motor. Armature voltage is fixed at its rated value.

relationship results in a parabolic curve whose peak occurs at one-half the no-load speed. Suppose our power requirement falls at point *B*, which is less than the peak power available but greater than the power available at that shaft speed. This is a common result and means a gear reduction is required to get more speed out of the motor and use its power capability more efficiently. Conversely, if our power requirement falls at point *A*, direct drive is a possibility.

3.5.1 Inertia Matching

An alternate method for motor selection is the matching of motor and load inertias to apply maximum acceleration to the load. Given the need for a gear reduction, we can go back to Figure 3.8(a) and assume the viscous friction is negligible. With a lubricated gear reducer, this assumption is incorrect; however, it does give us a design checkpoint, which can be changed if it does not meet our exact needs. For the equivalent system of Figure 3.8(c) we can write

$$nT_M = (J_L + n^2 J_M)\ddot{\theta}_L \tag{3.50}$$

[Doebelin, 1985] points out that since the torque effect on acceleration increases with *n*, while the inertial effect decreases acceleration as n^2, an optimum *n* should exist. It can be found by solving (3.50) for the load acceleration, differentiating with respect to *n*, and setting the result equal to zero for a maximum. The result is

$$n_{opt} = \sqrt{J_L/J_M} \tag{3.51}$$

This result is useful provided the rotor inertia for the motor we are considering is given in its data sheet. If it is not given (which is often the case for small motors), the rotor inertia can be roughly estimated given the overall size of the motor by using (3.14).

3.6 Mathematical Modeling of DC Motors and Transmissions

Dynamic mathematical models are now developed for the two most common cases where a DC motor drives an inertial load. These are direct drive and gear reduction drive and are depicted in Figure 3.14. We will see these models become powerful design tools when trying to predict the

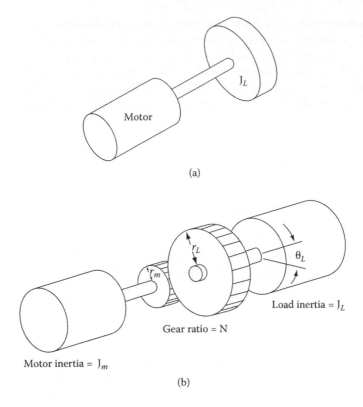

FIGURE 3.14
(a) Direct and (b) gear reduction drives.

transient behavior of a servo system. A consistent symbol set for the variables and constants involved is used for both models, and these symbols will be carried forward in succeeding chapters. The symbol set is given in Table 3.3.

In addition to the transfer function representation already used, an alternate modeling format called state space is introduced which is more amenable to computerized solution. Our goal is to be able to use either technique as needed to synthesize a solution to a particular problem. In general, we will find that velocity control systems involve second-order transfer functions and position control systems involve third-order transfer functions. To deal with transfer functions higher than second order, approaches other than the traditional time and frequency response methods introduced by Brown and others are needed. The state-variable approach in combination with a personal computer and appropriate software can fulfill this need.

TABLE 3.3

Variable and Constant Symbols List

Quantity	Symbol	Units
Input voltage	e_i	volt
Armature resistance	R_a, R	ohm
Armature inductance	L_a, L	henry
Back emf constant	K_e	volt/(rad/second)
Load angular speed	$\omega_o(t)$	rad/second
Motor angular speed	$\omega_i(t)$	rad/second
Armature current	$i_a(t)$	amp
Torque constant	K_T	Newton-meter/amp
Load damping coefficient	B_L	Newton-meter-second
Motor damping coefficient	B_M	Newton-meter-second
Load inertia	J_L	Newton-meter-sec^2
Motor rotor inertia	J_M	Newton-meter-sec^2
Load angular displacement	$\theta_o(t)$	radian
Motor angular displacement	$\theta_i(t)$	radian
Motor torque	$T_M(t)$	Newton-meter
Load torque	$T_L(t)$	Newton-meter
System equivalent damping	B_{eq}	Newton-meter-second
System equivalent inertia	J_{eq}	Newton-meter-sec^2
Gear ratio	n	dimensionless

3.7 Direct-Drive Model

DC motors drive a load directly in higher-speed, lower-torque applications. Speed control of the load is usually required—for example, the platen of a hard disk drive or the spindle of a CD player. Position control is less common in direct-drive systems, but for completeness models for position and speed control are developed.

3.7.1 Direct Drive—Transfer Function Representation

Including the armature inductance, we can write for the electrical circuit

$$i_a R_a + L_a \frac{di_a}{dt} + K_e \omega_o = e_i \tag{3.52}$$

where

$i_a \triangleq armature\ current, amp$

$R_a \triangleq armature\ resistance, ohm$

$L_a \triangleq$ *armature in ductance, henry*

$K_e \omega_o \triangleq$ *voltage drop due to back emf of motor, volt*

$K_e \triangleq$ *motor back emf constant,* $\dfrac{volt}{\dfrac{rad}{sec}}$

Next applying Newton's law to the motor shaft, we have

$$K_T i_a - B\omega_o = J\dot{\omega}_c \tag{3.53}$$

where

$K_T \triangleq$ *motor torque constant, Newton · meter/amp*

$B \triangleq$ *combined viscous damping of motor and load, N · m · sec*

$J \triangleq$ *combined inertia of motor and load, N · m · sec²*

In (3.52), the back emf voltage must oppose the voltage supplied by e_i. Similarly, the damping term of (3.53) opposes the torque produced due to the applied armature current. Invoking the D-operator, these two equations yield

$$i_a R_a + L_a D i_a + K_e \omega_o = e_i \tag{3.54}$$

$$K_T i_a - B\omega_o = J D \omega_o \tag{3.55}$$

Equations (3.54) and (3.55) constitute a model of the physical system shown in Figure 3.15. To write the transfer function between the input voltage and the output velocity, we will eliminate the armature current. From (3.54) we have

$$i_a = (e_i - K_e \omega_o)/(L_a D + R_a) \tag{3.56}$$

Substitution leads to

$$K_T (e_i - K_e \omega_o)/(L_a D + R_a) - B\omega_o = J D \omega_o \tag{3.57}$$

After some algebraic manipulation, the transfer function emerges as

$$\frac{\omega_o}{e_i}(D) = K_T / [(L_a D + R_a)(JD + B) + (K_T K_e)] \tag{3.58}$$

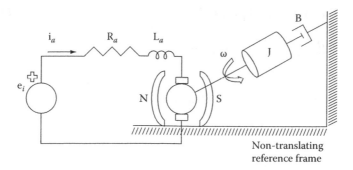

FIGURE 3.15
Direct drive physical system.

At this point it is useful to look at a block diagram according to [Kuo, 1982]. Figure 3.16 shows some elements of Equation (3.58) as separate entities included in a feedback loop. Using block diagram algebra, it can be shown that Figure 3.16 is equivalent to and represents (3.58). Although we are treating the DC motor as an open-loop system, it can be thought of as having a built-in feedback loop caused by the back emf. Physically, the back emf is proportional to the negative of the motor speed and constitutes an "electrical friction" that tends to improve the stability of the system in which the motor is used as an actuator. Figure 3.16 represents an alternative way of thinking about DC motor action, but for the remainder of the book it will be represented in its more "compact" form simply because the back emf is always present.

Now returning to our transfer function for a direct-drive system, multiplying through and gathering terms,

$$\frac{\omega_o}{e_i}(D) = K_T \,/\, [L_a J D^2 + (BL_a + R_a J)D + (R_a B + K_T K_e)] \qquad (3.59)$$

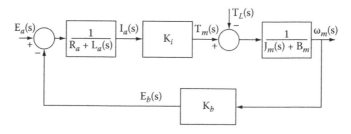

FIGURE 3.16
Block diagram of DC motor with internal "feedback loop."

In terms of the motor shaft position, we divide both sides of (5.8) by D and get

$$\frac{\theta_o}{e_i}(D) = K_T / [L_a J D^3 + (BL_a + R_a J)D^2 + (R_a B + K_T K_e)D] \qquad (3.60)$$

The above relation is in agreement with Kuo.

3.7.2 The State-Variable Approach to Dynamic Systems Modeling

The state-variable concept represents the so-called modern approach to describing the behavior of dynamic systems, whereas transfer functions represent the more traditional approach. Each has its place in control system design even today, and we will be using both approaches as needed in this book. The feature of the state-variable or "state-space" model description that sets it apart from the transfer function description is that the processes under examination are represented by systems of first-order differential equations. [Friedland, 1986] comments that the state-space representation is the most natural approach for a mathematician or physicist and that "were it not that much of classical control theory was developed by electrical engineers…, the state-space approach would have been in use much sooner." Since we plan to introduce MATLAB later in the book, we note that The MathWorks suggests the use of state space for higher accuracy in modeling calculations. On the other hand, [Ogata, 2008] uses transfer functions almost exclusively. We will be relying mostly on the transfer function representation in this book, but the reader is encouraged to be familiar with both methods.

According to Kuo, the state variables of a dynamic system are defined as "a minimal set of variables $x_1(t)$, $x_2(t)...x_n(t)$ such that knowledge of these variables at any time t_0, plus information on the input excitation subsequently applied, are sufficient to determine the state of the system at any time $t > t_0$." State variables are not necessarily system outputs which are measurable, but an output variable is usually defined as a function of the state variables.

3.7.3 Direct Drive—State-Variable Representation

The variables whose states we can define at $t = 0$, sufficient for defining the future state of the direct-drive DC servo system, are the load position θ_o, load velocity ω_o, and armature current i_a. We will represent the state of the system in the following form:

$$\frac{d}{dt}[x] = A[x] + B[u] \qquad (3.61a)$$

$$y(t) = C[x] + D[u] \qquad (3.61b)$$

where A, B, C, and D are matrices of appropriate dimensions, and

$$x \triangleq \text{state vector} \tag{3.61c}$$

$$u \triangleq \text{input vector} \tag{3.61d}$$

$$y \triangleq \text{output vector} \tag{3.61e}$$

For a system with a controlled velocity output, i_a and ω_o are the two system states, e_i is the input, and ω_o is the output. State 1, the induced armature current, can be found by rearranging (3.54) in terms of the current yielding

$$\frac{d}{dt} i_a = -\frac{R_a}{L_a} i_a - \frac{K_s}{L_a} \omega_o + \frac{1}{L_a} e_i \tag{3.62}$$

State 2, the resulting angular rate, is found by a similar rearrangement of (3.55), giving

$$\frac{d}{dt} \omega_o = -\frac{B}{J} \omega_o - \frac{K_T}{J} i_a \tag{3.63}$$

For direct-drive velocity output control, the [A] matrix in (3.61a) is a 2×2 matrix. The state vector is a 1×2 matrix given by

$$\frac{d}{dt} \begin{bmatrix} i_a \\ \omega_o \end{bmatrix} = \begin{bmatrix} \frac{-R_a}{L_a} & \frac{-K_s}{L_a} \\ \frac{K_T}{J} & -\frac{B}{J} \end{bmatrix} \cdot \begin{bmatrix} i_a \\ \omega_o \end{bmatrix} + \begin{bmatrix} 1/L_a \\ 0 \end{bmatrix} \cdot e_i \tag{3.64}$$

The corresponding output vector is

$$y(t) = [0\ 1] \cdot \begin{bmatrix} i_a \\ \omega_o \end{bmatrix} + [0] \cdot e_i \tag{3.65}$$

For position control, a third state can be added as

$$\frac{d}{dt} \theta_o = \omega_o \tag{3.66}$$

For direct-drive position output control, the [A] matrix in (3.61a) is a 3×3 matrix. The state vector is a 1×3 matrix given by

$$\frac{d}{dt} \begin{bmatrix} i_a \\ \omega_o \\ \theta_o \end{bmatrix} = \begin{bmatrix} \frac{-R_a}{L_a} & \frac{-K_e}{L_a} & 0 \\ \frac{K_T}{J} & -\frac{B}{J} & 0 \\ 0 & 1 & 0 \end{bmatrix} \cdot \begin{bmatrix} i_a \\ \omega_o \\ \theta_o \end{bmatrix} + \begin{bmatrix} 1/L_a \\ 0 \\ 0 \end{bmatrix} \cdot e_i \tag{3.67}$$

The corresponding output vector is

$$y(t) = [0\ 0\ 1] \cdot \begin{bmatrix} i_a \\ \omega_o \\ \theta_o \end{bmatrix} + [0] \cdot e_i \tag{3.68}$$

3.8 Motor and Gear Train Model

DC motors drive loads through gear trains in lower-speed, higher-torque applications. Position control of the load is usually required—for example, the print cartridge carriage mechanism of an ink-jet printer or the various axes of an industrial robot. Speed control is less common in gear reduction drive systems, but for completeness models for position and speed control are developed.

3.8.1 Gear Train Drive—Transfer Function Representation

Recalling the electrical circuit equation for a direct-drive system, we had

$$i_a R_a + L_a \frac{di_a}{dt} + K_e \omega_o = e_i \tag{3.69}$$

We now insert a "reduction" gear train between motor and load, so called because the output velocity is lower than the input velocity. The two are related by the gear ratio n such that

$$\omega_o = \omega_i / n \tag{3.70}$$

Since the back emf constant of the motor is related to the motor's shaft speed, (3.69) must now be modified so that for the motor shaft we have

$$i_a R_a + L_a \frac{di_a}{dt} + K_e \omega_i = e_i \tag{3.71}$$

In terms of the output speed, by substitution we get

$$i_a R_a + L_a \frac{di_a}{dt} + K_e n \omega_o = e_i \tag{3.72}$$

Similarly, for the direct-drive system we had for the motor shaft

$$K_T i_a - B\omega_o = J\dot{\omega}_o \qquad (3.73)$$

With the reduction gear train inserted, the torque supplied by the motor shaft is multiplied by the gear ratio, and we are trading off torque for speed so that

$$T_{o(app)} = nT_M = nK_T i_a \qquad (3.74)$$

where

$$T_{o(app)} \triangleq \text{torque applied to load by output gear}$$

$$T_M \triangleq \text{input torque supplied by motor}$$

We also recall that for a geared system, the equivalent damping coefficient and inertia referred to the load are

$$\text{equivalent damping} \triangleq B_{eq2} = n^2 B_M + B_L \qquad (3.30a)$$

$$\text{equivalent inertia} \triangleq J_{eq2} = n^2 J_M + J_L \qquad (3.30b)$$

Substituting all of this into (3.73) gives

$$nK_T i_a - B_{eq}\omega_o = J_{eq}\dot{\omega}_o \qquad (3.75)$$

If we invoke the D-operator again, the two equations (3.72) and (3.75) yield

$$i_a R_a + L_a D i_a + nK_e \omega_o = e_i \qquad (3.76)$$

$$nK_T i_a - B_{eq}\omega_o = J_{eq}D\omega_o \qquad (3.77)$$

Equations (3.76) and (3.77) constitute a model of the physical system shown in Figure 3.17. Again after we eliminate the armature current and gather terms, the transfer function is

$$\frac{\omega_o}{e_i}(D) = nK_T / [L_a J_{eq} D^2 + (B_{eq}L_a + R_a J_{eq})D + (R_a B_{eq} + n^2 K_T K_e)] \qquad (3.78)$$

In terms of the load shaft position, we divide both sides of (3.78) by D to get

$$\frac{\theta_o}{e_i}(D) = nK_T / [L_a J_{eq} D^3 + (B_{eq}L_a + R_a J_{eq})D^2 + (R_a B_{eq} + n^2 K_T K_e)D] \qquad (3.79)$$

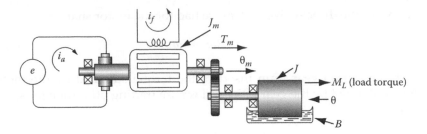

FIGURE 3.17
Gear reduction drive physical system.

3.8.2 Gear Reduction Drive—State-Variable Representation

The variables whose states we can define at $t = 0$, sufficient for defining the future state of the gear reduction drive DC servo system, are the load position θ_o, load velocity ω_o, and armature current i_a. For a system with a controlled velocity output, i_a and ω_o are the two system states, e_i is the input, and ω_o is the output. State 1, the induced armature current, can be found by rearranging (3.76) in terms of the current, yielding

$$\frac{d}{dt}i_a = -\frac{R_a}{L_a}i_a - \frac{nK_e}{L_a}\omega_o + \frac{1}{L_a}e_i \tag{3.80}$$

State 2, the resulting angular rate, is found by a similar rearrangement of (3.77), giving

$$\frac{d}{dt}\omega_o = -\frac{B_{eq}}{J_{eq}}\omega_o - \frac{nK_T}{J_{eq}}i_a \tag{3.81}$$

For gear-reduction drive velocity output control, the [A] matrix is a 2×2 matrix. The state vector is

$$\frac{d}{dt}\begin{bmatrix} i_a \\ \omega_o \end{bmatrix} = \begin{bmatrix} \frac{-R_a}{L_a} & \frac{-nK_e}{L_a} \\ \frac{nK_T}{J_{eq}} & \frac{-B_{eq}}{J_{eq}} \end{bmatrix} \cdot \begin{bmatrix} i_a \\ \omega_o \end{bmatrix} + \begin{bmatrix} 1/L_a \\ 0 \end{bmatrix} \cdot e_i \tag{3.82}$$

The corresponding output vector is

$$y(t) = [0 \ 1] \cdot \begin{bmatrix} i_a \\ \omega_o \end{bmatrix} + [0] \cdot e_i \tag{3.83}$$

For position control, a third state is added as before,

$$\frac{d}{dt}\theta_o = \omega_o \qquad (3.84)$$

For gear-reduction drive position output control, the $[A]$ matrix is a 3×3 matrix. The state vector is

$$\frac{d}{dt}\begin{bmatrix} i_a \\ \omega_o \\ \theta_o \end{bmatrix} = \begin{bmatrix} \frac{-R_a}{L_a} & \frac{-nK_e}{L_a} & 0 \\ \frac{nK_T}{J_{eq}} & -\frac{B_{eq}}{J_{eq}} & 0 \\ 0 & 1 & 0 \end{bmatrix} \cdot \begin{bmatrix} i_a \\ \omega_o \\ \theta_o \end{bmatrix} + \begin{bmatrix} 1/L_a \\ 0 \\ 0 \end{bmatrix} \cdot e_i \qquad (3.85)$$

The corresponding output vector is

$$y(t) = [0\ 0\ 1] \cdot \begin{bmatrix} i_a \\ \omega_o \\ \theta_o \end{bmatrix} + [0] \cdot e_i \qquad (3.86)$$

References

Brown, G., and D. Campbell, *Principles of Servomechanisms*, Wiley, 1938.
Chapman, S., *Electric Machinery Fundamentals*, WCB McGraw-Hill, 1999.
Doebelin, E., *Control System Principles and Design*, Wiley, 1985.
Doebelin, E., *Dynamic Analysis and Feedback Control*, McGraw-Hill, 1962.
Doebelin, E., *System Dynamics: Modeling and Response*, Merrill, 1972.
Electro-Craft Corp., *DC Motors, Speed Controls, Servo Systems*, 5th ed., 1980.
Friedland, B., *Control System Design: An Introduction to State-Space Methods*, Dover, 1986.
Kuo, B., and J. Tal, *Incremental Motion Control: DC Motors and Control Systems*, SRL, 1978.
Kuo, B., *Automatic Control Systems*, Prentice-Hall, 1982.
Ogata, K., *Matlab for Control Engineers*, Pearson Prentice-Hall, 2008.
Stiffler, K., *Design with Microprocessors for Mechanical Engineers*, McGraw-Hill, 1992.

4

Feedback Control Systems

4.1 Introduction

There are several advantages to the use of feedback in control systems, as illustrated by the example of a velocity servo in Chapter 1. However, there are also potential problems with feedback systems that must be mitigated in their design. Perhaps the most important of these is the natural tendency of systems with feedback to oscillate, just like any other vibrating system. The unique problem with feedback systems is that the magnitude of their oscillations can increase without bound if they are not properly designed. We will first take a brief detour into the realm of vibrating systems and see in some detail how they are analyzed mathematically. Then, we will relate the analysis technique to closed-loop feedback systems.

4.2 Mathematical Notation

The differential equations presented so far in this text have used notation involving the D-operator, that is,

$$D \triangleq \frac{d}{dt} \tag{4.1}$$

This notation represents a kind of introduction or "subset" of a much larger selection of analysis techniques. Any treatment of feedback control systems cannot be considered complete without presenting this larger tool set, at least at an introductory level. In the following sections, we will show why the D-operator works and its relationship to more rigorous techniques.

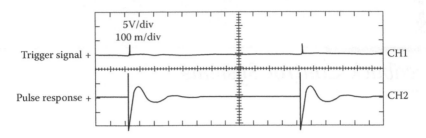

FIGURE 4.1
Pulse response of a closed-loop control system.

4.3 Linear, Time-Invariant Systems

The differential equations we are using to describe a physical system to be controlled, or already under control, have the properties of linearity and time-invariance. A linear system is one where if the input consists of the weighted sum of several signals, the output is simply the superposition, or weighted sum, of the responses of the system to each of those signals. A system is time-invariant if a time shift in the input signal causes an equal time shift in the output signal. Time-invariance implies that the coefficients of our equations, such as damping, inertia, and inductance, remain constant over the time period under investigation. For example, it is well known in control system practice that many closed-loop systems exhibit a fast exponential rise, followed by a damped, oscillatory response to an abrupt input (i.e., a square pulse input), as shown in Figure 4.1. For the system to be linear, its input signal would have to be the weighted sum of components whose individual system responses make up the output signal of Figure 4.1. It can be shown that a square pulse input can be resolved into the sinusoidal components of a Fourier series. We will see that both an exponential output response and a damped sinusoidal output response are the result of a linear combination of all the sinusoidal components of the input pulse.

4.4 Oscillations, Rotating Vectors, and the Complex Plane

Fortunately, mathematicians discovered a method of describing both exponential responses and sinusoidal responses using the same form of mathematical function. Their motivation was twofold: (1) these responses are seen in combination with each other so frequently in nature, and (2) exponential functions are much easier to manipulate algebraically. Essentially,

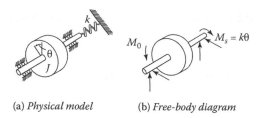

(a) *Physical model*   (b) *Free-body diagram*

FIGURE 4.2
Rotor of a torque motor, (a) physical model and (b) free-body diagram.

the problem was to represent the sinusoidal oscillation of a system, such as
the one shown in Figure 4.2, in exponential form. This device is the rotor
of a "torque motor," somewhat similar to a DC motor except that the rotor
is restricted to small movements by a torsional spring. Such a device might
be used in place of a solenoid to position a valve more precisely. It is not a
DC servo but can be thought of as a precursor to it, to aid in understanding
how system responses relate to the complex plane. The undamped sinusoidal
oscillation of this second-order system is

$$\theta = \theta_{max} \cos(\omega t + \varphi) \tag{4.2}$$

where

$$\omega = \sqrt{k/J} \tag{4.2a}$$

It can be shown [Cannon, 1967] that this motion can be represented in
the form

$$\theta = Ce^{st} \tag{4.3}$$

by considering a new variable s to be a complex number. The basis for the
manipulation of complex numbers is Euler's equation:

$$e^{i\phi} = \cos\phi + i\sin\phi \tag{4.4}$$

where the imaginary part of this equation is multiplied by

$$i = \sqrt{-1} \tag{4.4a}$$

We also note that [Reswick and Taft, 1967]

$$-i = \sqrt{-1} \tag{4.4b}$$

We now proceed to prove that (4.3) is indeed equivalent to (4.2). We have

$$\theta = Ce^{st} = C_1 e^{i\omega t} + C_2 e^{-i\omega t} \tag{4.5}$$

$$\theta = (C_1 + C_2)\cos \omega t + i(C_1 - C_2)\sin \omega t \tag{4.5a}$$

or

$$\theta = C_3 \cos(\omega t - \varphi) \tag{4.5b}$$

In this case of free oscillation the complex variable s has taken the form

$$s = \pm i\omega \tag{4.6}$$

As Cannon explains, "considerable convenience and saving of labor can be realized by using a rotating arrow, instead of a time plot, to represent a sinusoidally varying quantity." The arrow rotation is accomplished graphically by the use of complex numbers with "imaginary" parts. This way the sinusoidal plot can stay stationary on the printed page, in the complex plane. Because this is done so routinely in control work, it is important to demonstrate it rigorously. We first draw the complex vector $e^{i(\omega t - \varphi)}$ in the complex plane as shown in Figure 4.3a. Imagine that this vector rotates counterclockwise with angular velocity ω. Then we draw the complex conjugate vector $e^{-i(\omega t - \varphi)}$ and

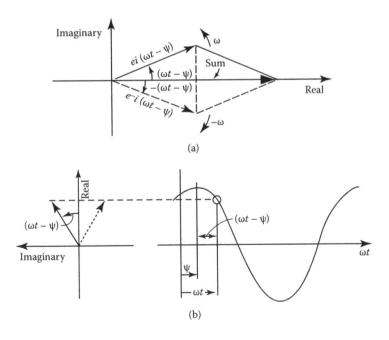

(a)

(b)

FIGURE 4.3
Rotating complex vectors, (a) conjugate pair and (b) sine wave representation.

imagine it rotating clockwise with angular velocity $-\omega$. When these equations are put into the form of (4.4) and added, the result is a real quantity:

$$e^{i(\omega t-\varphi)} + e^{-i(\omega t-\varphi)} = 2\cos(\omega t - \varphi) \tag{4.7}$$

That is, the imaginary components cancel out, leaving only the real components, which grow and shrink along the real axis sinusoidally with time. From this we can conclude that a cosine wave can be represented in the exponential form

$$A\cos(\omega t - \varphi) = \frac{1}{2} \, Re \, A[e^{i(\omega t-\varphi)} + e^{-i(\omega t-\varphi)}] \tag{4.8}$$

Because each exponential component of (4.7) has the same real part, we can eliminate the conjugate for convenience and say

$$A\cos(\omega t - \varphi) = ReA[e^{i(\omega t-\varphi)}] \tag{4.9}$$

Figure 4.3b shows the meaning of the quantity φ; it is a constant that allows the expression $\cos(\omega t - \varphi)$ to represent a cosine wave having its maximum at an arbitrary time given by

$$t = \varphi/\omega \tag{4.9a}$$

In the rotating vector picture, $-\varphi$ is the angle of the vector at $t = 0$ and φ is known as the phase angle of the vector. A rotating vector with a phase angle is also called a phasor (a concatenation of the words *phase* and *vector*).

4.5 From Fourier Series to Laplace Transform

We now propose to evolve the Fourier series of an input motion time function into its Laplace transform. The reason behind this evolution is to enable us to treat the entire range of input motions mathematically from periodic to nonperiodic and from continuous to abrupt. A common example is a "step input" function of motion, which the Laplace transform method can accommodate with relative ease. Our ultimate goal is to be able to express almost any arbitrary function as an exponential in the form of (4.3). [Cannon, 1967] summarizes the value of the Laplace transform this way:

> It transforms the routine mathematical part of a [dynamics] problem to another domain, where operations and manipulations are much easier to perform, then transports it back after the work is done.

Figure 4.4 shows a range of possible input motion functions we would like to be able to accommodate with a single analytical method. Figure 4.5 shows

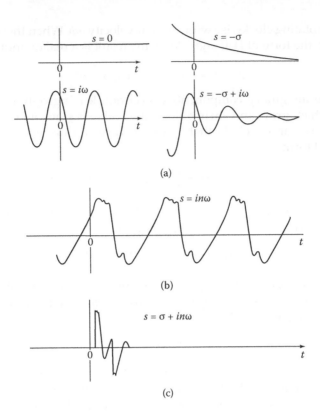

FIGURE 4.4
Range of possible input motion functions to be expressed in the form Ce^{st}.

the progression by which we will synthesize the Laplace transform of a function from its Fourier series. This progression closely follows Cannon and [Gardner and Barnes, 1956].

The strategy starts by taking an arbitrary, nonperiodic function and pretending that it is periodic, as shown in Figure 4.6. We then represent this by a Fourier series, letting its period grow to infinity and seeing how the series is affected. The periodic function of Figure 4.6 can be expressed as

$$x(t) = \left[\frac{1}{2\pi} \sum_{n=-\infty}^{\infty} \omega_0 C(s) e^{st} \right], \text{ where } s = in\omega_0 \qquad (4.10a)$$

where the coefficients of (4.10a) are

$$C(s) = \left[\int_{-\pi/\omega_0}^{\pi/\omega_0} x(t) e^{-st} \, dt \right], \text{ where } s = in\omega_0 \qquad (4.10b)$$

Evolution of the Laplace Transform

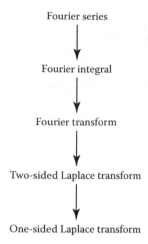

FIGURE 4.5
Mathematical evolution of the Laplace transform.

and

$$\omega_0 = 2\pi/T \tag{4.10c}$$

Letting the period T of Figure 4.6 and (4.10) grow to infinity, the separation ω_0 between succeeding terms in the series approaches zero, and in the limit we get the Fourier integral

$$x(t) = \left[\frac{1}{2\pi} \int_{s=-\infty}^{\infty} ds\, C(s)e^{st} \right], \quad \text{where} \quad s = i\omega \tag{4.11a}$$

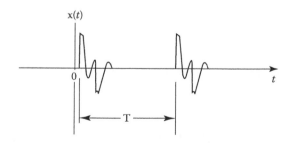

FIGURE 4.6
Letting a nonperiodic function become periodic with variable period T.

and its spectrum, the Fourier transform:

$$C(s) = \left[\int_{t=-\infty}^{\infty} x(t)e^{-st}\,dt \right], \quad \text{where} \quad s = i\omega \tag{4.11b}$$

To ensure convergence [Cannon, 1967] for a wide class of functions, including the step function, we modify the integral in (4.11b) by letting $s = \sigma + i\omega$. The result is the "two-sided" Laplace transform of $x(t)$:

$$\mathcal{L}[x(t)] = X(s) \triangleq \int_{t=-\infty}^{\infty} x(t)e^{-st}\,dt \tag{4.12}$$

4.6 Elementary Laplace Transforms

In the analysis of linear DC servos, we are mostly concerned with the derivatives and integrals of time functions. For the derivative of a function, it can be shown that

$$\mathcal{L}\left[\frac{dx}{dt}\right] = sX(s) \tag{4.13}$$

The expression (4.13) is known as a Laplace transform pair. For the integral of a function, the Laplace pair is

$$\mathcal{L}\left[\int x(t)dt\right] = \frac{X(s)}{s} \tag{4.14}$$

For higher-order derivatives, the Laplace pair is

$$\mathcal{L}\left[\frac{d^n x}{dt^n}\right] = s^n X(S) \tag{4.15}$$

For higher-order integrals, the Laplace transform for the nth integral of $x(t)$ is

$$X(s)/s^n \tag{4.16}$$

A simple pattern is formed by (4.13) to (4.16), such that $X(s)$ is multiplied by s^n for a derivative and divided by s^n for an integral. If the letter D (for *derivative*) is substituted for s, we travel full circle and arrive at the "D-operator" used earlier in this book. The *D*-operator is typically used for simplicity, where students have not been yet exposed to the Laplace transform. The reader is cautioned that this shortcut only works for the straightforward cases of derivatives and integrals. Other Laplace transform pairs are tabulated for reference in the appendices of [Cannon, 1967] and [Kuo, 1982].

4.7 System Analysis Using Laplace Transforms

Returning to Figure 4.2, the equation of motion for this undamped system is

$$J\ddot{\theta} + k\theta = 0 \tag{4.17}$$

If a sudden step input torque is applied to this system, by electromagnetic means, we have

$$J\ddot{\theta} + k\theta = M_0 u(t) \tag{4.18}$$

Because we are only interested in motion starting at $t = 0$, we choose to transform (4.18) to the *s*-domain using the "one-sided" [Cannon, 1967] Laplace transform. The result is

$$J[s^2\theta(s) - s\theta(0-) - \dot{\theta}(0-)] + k\theta(s) = M(s) = M_0/s \tag{4.19}$$

The second and third terms inside the brackets denote the values of the displacement and velocity of the rotor immediately before $t = 0$. The transfer function between M and θ is obtained directly from (4.19), with initial velocity and displacement equal to zero. The result is

$$\frac{\theta(s)}{M(s)} = \frac{1}{Js^2 + k} \tag{4.20}$$

Note that the transform of the input moment is not used in deriving the transfer function, since the transfer function represents the system only and not its input. However, it will be used to obtain the complete response of the system. This is accomplished by first solving (4.19) for the motion $\theta(s)$. After some routine algebra, we get

$$\theta(s) = \frac{M_o}{J}\left[\frac{1}{s\left(s^2 + \frac{k}{J}\right)}\right] \tag{4.21}$$

This expression (4.21) is the system response function. The system's natural behavior can be studied by taking the denominator of (4.20) or (4.21) and setting it equal to zero. We have

$$s\left[s^2 + \frac{k}{J}\right] = 0 \qquad (4.22)$$

This expression (4.22) is known as the characteristic equation of the system. We have one root of this equation at the origin of the complex plane and another pair of roots that are purely imaginary, that is,

$$s = 0 \qquad (4.23a)$$

$$s = \sqrt{-\frac{k}{J}} = \pm i\sqrt{k/J} \qquad (4.23b)$$

These roots are plotted on the complex plane in Figure 4.7. Substituting the complex pair of roots into an assumed exponential response, we get a purely sinusoidal natural behavior:

$$\theta = e^{st} = e^{\pm i\sqrt{k/J}t} = e^{\pm i\omega t} \qquad (4.24)$$

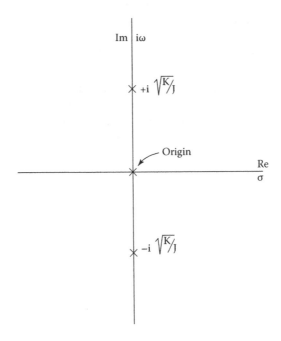

FIGURE 4.7
Three roots of vibrating system plotted in the complex plane.

where

$$\omega = \sqrt{k/J} \qquad (4.2a)$$

Substituting the root at the origin, we get a DC level at a magnitude of unity. Our final step is to inverse transform the system response function to see what the response is as a function of time. We expect purely sinusoidal behavior combined with a DC level. Inverse transformation can be done by using either a table of transforms or partial fraction expansion. Looking at (4.21), we see that the denominator is third order. Referring to Cannon's transform table, we see that for

$$F(s) = 1/\left[s(s^2 + \omega_a^2)\right] \qquad (4.25)$$

$$F(t) = \left(\frac{1}{\omega_a^2}\right)[1 - \cos(\omega_a t)] \qquad (4.26)$$

Using this transform pair, the time response is

$$\theta(t) = \left(\frac{M_o}{k}\right)[1 - \cos(\omega t)] \qquad (4.27)$$

This response is plotted in Figure 4.8. Its physical interpretation: The moment is applied suddenly, bringing the position of the rotor to a new equilibrium position, balanced by the spring constant of the torsional spring. Because no damping was assumed, we have a sustained oscillation about the new equilibrium position. In practice, the spring itself will exhibit some damping action due to its own internal heating, and the rotor would eventually stop oscillating, slowly decaying in an exponential fashion.

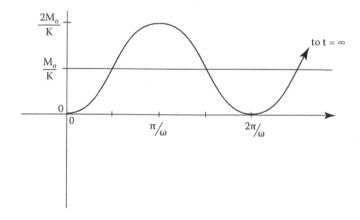

FIGURE 4.8
Time response of vibrating system to a step input.

4.7.1 Final and Initial Value Theorems

Two special cases of the preceding derivation can serve to give a quick initial view of system behavior. The final value theorem states that

$$x(\infty) = \lim_{t \to \infty} x(t) = \lim_{s \to 0}[sX(s)] \tag{4.28}$$

That is, the time response of a variable $x(t)$ after a long time can be obtained by multiplying its response function by s and evaluating it as s approaches zero (provided that all roots of the characteristic equation are in the left half of the s-plane or at the origin). Similarly, the initial value theorem states that

$$x(0+) = \lim_{t \to 0+} x(t) = \lim_{s \to \infty}[sX(s)] \tag{4.29}$$

That is, the time response of a variable $x(t)$ at a time right after $t = 0$ can be obtained by multiplying its response function by s and evaluating it as s approaches infinity. These theorems are stated without proof, and their derivations can be found in [Gardner and Barnes, 1956] and elsewhere. For our preceding example, the system response function (4.21) multiplied by s gives

$$s\theta(s) = \frac{M_o}{J}\left[\frac{1}{\left(s^2 + \frac{k}{J}\right)}\right] \tag{4.30}$$

The final value is taken as s approaches zero. The expression reduces to

$$\theta(\infty) = \frac{M_o}{k} \tag{4.31a}$$

The initial value is taken as s approaches infinity. The expression reduces to

$$\theta(0+) = 0 \tag{4.31b}$$

4.8 Philosophy of Feedback Control

We now have some necessary tools at our disposal to look at a new class of systems employing the concept of feedback. What is feedback, and why is it desirable? As introduced in Chapter 1, some advantages are realized in the DC velocity servo if we do the following:

1. Measure what the speed actually is
2. Compare this value to a desired value
3. Adjust the driving voltage to reduce the error if one exists

The principal advantages of this scheme are as follows:

1. The ability to control motions with acceptably small error (i.e., high accuracy)
2. Improvement in speed of response to input changes or disturbances
3. Performance insensitivity to changes in physical parameters

Feedback control essentially involves a trade-off between these potential performance improvements and the risk of instability or sustained oscillation. Specifically, the two major problems we will discuss here are accuracy and stability, which naturally conflict with each other.

4.8.1 Terminology of Loop Closing

In a feedback system, a feedback "loop" is created from the output (controlled variable) back to a controller that compares it with the input (desired value). Figure 4.9 shows the basic feedback loop in general form. Quantity V is the input, C is the output of interest, and E is the difference signal produced at the feedback summing junction S1. We note that these quantities V, C, and E are the Laplace transforms of time functions $v(t)$, $c(t)$, and $e(t)$. The function G is the product of all transfer functions in the forward path from E to C. Similarly, the function H is the product of all transfer functions in the feedback path from C to the summing junction. The closed-loop transfer function of the entire system can be derived using the relation

$$C = GE = G(V - HC) \tag{4.32}$$

After some routine algebra,

$$\frac{C}{V} = \frac{G}{1+GH} \tag{4.33}$$

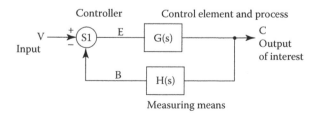

FIGURE 4.9
Basic feedback control system block diagram.

The quantity $\frac{B}{E} = GH$ is known as the open-loop transfer function. The characteristic equation of the feedback loop is given by setting the denominator of (4.33) equal to zero, that is,

$$1 + GH = 0 \tag{4.34}$$

The denominator itself $1 + GH$ is known as the characteristic function of the feedback loop.

4.9 Accuracy of Feedback Systems

For perfect accuracy, we would like the transfer function C/V to be unity. By inspection of (4.33), we see that using feedback allows the designer to approach this ideal but never achieve it. For a fixed value of feedback gain H, as the forward gain is increased, the system accuracy improves. For relatively high values of G, a system may respond to a disturbance as shown in Figure 4.10. Two types of errors occur in general: (1) a transient error and (2) a steady-state error. In feedback systems, an error must occur before corrective action is taken. After corrective action, the steady-state error persists for all time, preventing perfect accuracy. Because no system can achieve perfect accuracy, the steady-state error is made small enough by the designer to both satisfy system specifications and outperform a similarly designed open-loop system.

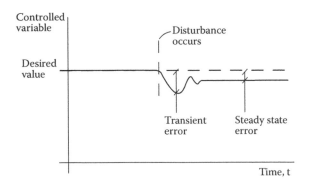

FIGURE 4.10
Transient and steady-state errors in a control system after a disturbance.

4.10 Stability of Feedback Systems

Broadly speaking, each component of a control system has a certain time lag effect associated with it. These lags are usually due to first-order effects (i.e., exponential time constant delays), but we assume the lag times are definite (known as "dead times") for simplicity. Following [Doebelin, 1962], Figure 4.11 shows the general layout of a feedback control system, and Figure 4.12 is a time-plot of various points in the system at a given time. Suppose the system's requirement is to keep the controlled variable within a certain tolerance of a desired value, and the process has been disturbed at time t_0. The controlled variable has moved off to the higher side of where it was intended to be. The task of the controller and final control element is to oppose this trend and bring the controlled variable back within specification. The various lags in the system prevent this from happening until time t_1 later. If we continue to observe the system, the controlled variable is moving toward the desired value as needed. However, because of time lags the final control element does not receive this news until it is too late. At time t_2, the final control element is acting to aid the downward trend, even though the controlled variable has already reached its target value. This causes the controlled variable to be forced downward even further. The controller action is said to be "out of phase" with the variable being controlled. One important design variable is the sensitivity or gain of the controller, that is, the corrective effort per unit of error. If the sensitivity of the controller is high enough, the out-of-phase behavior can lead to sustained oscillations known as absolute instability. Sustained oscillations in a DC position servo are not desirable and can cause physical damage to the system.

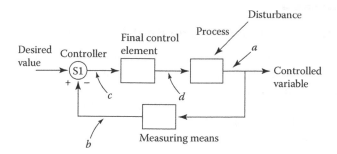

FIGURE 4.11
General layout of a feedback control system to examine time lag effects.

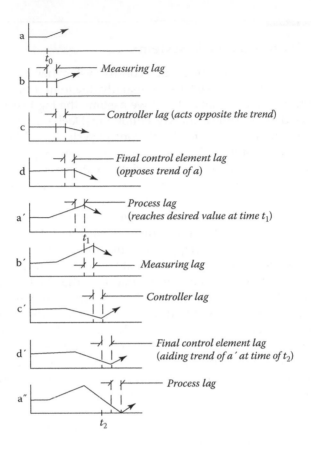

FIGURE 4.12
Time plots at various points in the control system of Figure 4.11.

4.11 Stability Assessment—The Root-Locus Method

A powerful method of assessing the relative stability of control systems as parameters are varied was invented by Evans in 1949. Despite its age, the root-locus method is widely employed today and is an integral part of the MATLAB control system toolbox. We will demonstrate the method now using a third-order position control servo.

Figure 4.13 shows a position control system in both physical model and block diagram form. From (4.34), we know that the system's characteristic equation is of the general form

$$1 + GH = 0 \qquad\qquad (4.35)$$

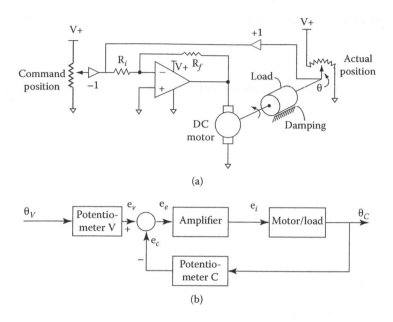

(a)

(b)

FIGURE 4.13
Placement of DC motor inside a single-loop servo, (a) physical model and (b) block diagram.

where GH is the system's open-loop transfer function (also known as the "loop gain"). Going around the loop, we can say

$$GH = \frac{e_i}{e_e}(s) \times \frac{\theta_c}{e_i}(s) \times \frac{e_c}{\theta_c}(s) \tag{4.36}$$

We define the amplifier gain as a variable, that is,

$$\frac{e_i}{e_e}(s) = K_A \tag{4.37}$$

For simplicity, suppose the potentiometer gain factor is unity. The motor/load transfer function is taken directly from Chapter 3, substituting s for the D-operator:

$$\frac{\theta_c}{e_i}(s) = K_T / [LJs^3 + (BL + RJ)s^2 + (RB + K_T K_e)s] \tag{4.38}$$

In this relation, the variable R is the motor's armature resistance, the variable L is the motor's armature inductance, and the subscript a has been dropped. The open-loop transfer function becomes

$$GH = K_A K_T / [LJs^3 + (BL + RJ) s^2 + (RB + K_T K_e)s] \tag{4.39}$$

The characteristic equation of (4.35) may be written as [Ogata, 2008]

$$1 + K\frac{num}{den} = 0 \qquad (4.40)$$

where *num* and *den* are the numerator and denominator polynomials of the loop gain, *GH*, and *K* is the amplifier gain, which is varied from zero to infinity. We can now use MATLAB's control system toolbox to plot the root loci of the DC motor and amplifier inside a position control loop, using the following code:

```
% DC motor
R = 2.0; % ohm
L = 0.5; % henry
Kt = .015; % N-m/A
Ke = .015; % V-s/rad
J = .02; % kg-m^2
B = 0.2; % N-m-s
num5 = Kt;
den5 = [J*L (J*R+B*L) (Kt*Ke+R*B) 0];
sys_dc5 = tf(num5,den5);
```

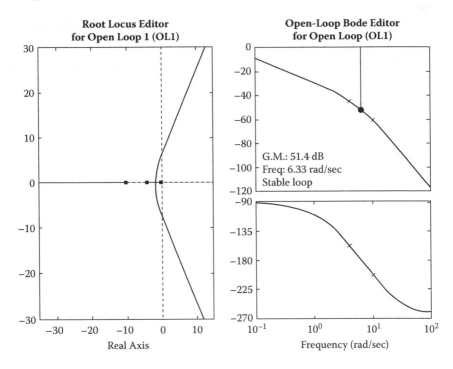

FIGURE 4.14
(a) Root locus, (b) magnitude, and (c) phase shift frequency response plots of a third-order, single-loop DC servo.

Figure 4.14 shows the three-part output of the "sisotool(sys_dc5)" statement at the MATLAB command line. Figure 4.14(a) shows the root locus plot for our third-order DC servo. Three points are plotted with the "■" symbol, to remind us that these roots are the poles of the closed-loop transfer function of our control system in Figure 4.13. As the gain K_A is increased, these three roots together follow specific trajectories in the s-plane. The leftmost "electrical pole" moves along the negative real axis toward $-\infty$. The middle "mechanical pole" and the pole at the origin move toward each other, meet, and then depart along symmetrical paths above and below the real axis. The system becomes marginally stable when any part of the trajectory crosses the imaginary axis and into the right half-plane.

Figures 4.14(b) and (c) show the gain magnitude and phase plots ("Bode" plots) for the frequency response of this servo. Essentially, the servo behaves as a low-pass filter. The gain margin is the difference between the 0 dB magnitude and the magnitude at a phase shift of $-180°$. The phase margin is the difference between $-180°$ and the phase shift at 0 dB. A gain of about 370 (51.4 dB) can be tolerated at a frequency of 6.33 rad/sec before oscillation occurs.

References

Cannon, R., *Dynamics of Physical Systems*, McGraw-Hill, 1967.
Doebelin, E., *Dynamic Analysis and Feedback Control*, McGraw-Hill, 1962.
Gardner, M., and J. Barnes, *Transients in Linear Systems*, Wiley, 1956.
Kuo, B., *Automatic Control Systems*, Prentice-Hall, 1982.
Ogata, K., *Matlab for Control Engineers*, Pearson Prentice-Hall, 2008.
Reswick, J., and C. Taft, *Introduction to Dynamic Systems*, Prentice-Hall, 1967.

5

Proportional Control of a
Second-Order DC Servo

5.1 Introduction

We have already seen that a DC position servo can be modeled by the use of third-order differential equations. A second-order model of this system is only an approximation of what is really happening, so why explore it? The simple answer is that control engineers communicate via the "language" of the second-order system. Statements like "You need more damping" or "There is too much gain" are very common in control work. These statements must be taken seriously, because a third-order system behaves much like one of second order, provided that two roots of the characteristic equation "dominate" the system response to standard test inputs. The dominant roots are those that are closest to the origin of the s-plane, along the negative real axis. Concentrating attention on these two roots can give the control engineer an intuitive feeling for how the system behaves. Fortunately, the second-order system has been well characterized and lends itself to manual solution using preplotted graphs, charts, and Laplace transform tables. We now proceed to investigate a second-order approximation of a DC position servo. Our discussion is limited to proportional control only, so that results are easily compared when the system is subjected to step, ramp, and sinusoidal inputs.

5.2 Proportional Control

After on-off or "bang-bang" control, proportional control is the next logical step of control refinement. It applies a corrective action to a plant which is proportional to the magnitude of and has the same sign as the error signal. As shown in Figure 5.1, a linear amplifier is inserted between the summing junction and the plant of a closed-loop control system. In our case, the plant consists of a DC motor, inertial load, and gearing. For simplicity, the transfer

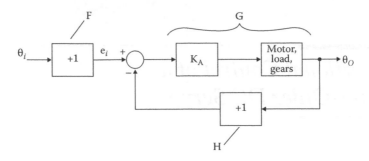

FIGURE 5.1
Second-order closed-loop position servo with proportional control.

functions from both positions to voltage are assumed to be unity, that is, 1 V/ rad. The gain K_A of the amplifier can be varied, usually with a potentiometer within its own feedback circuitry.

5.3 Second-Order Approximation

Ignoring one energy storage element, the armature inductance, is the basic tenet of the traditional approach, which we will take here. The assumption is that the electrical dynamics will occur in approximately one-tenth the time taken by the mechanical dynamics. This means that the inertia of the drive train is the dominant energy storage element, and electrical transients disappear ten times faster. This assumption gives a first-order system representing the mechanical damping and inertia, and a pure integration in cascade representing the integrating nature of the motor from velocity to position. The result is a second-order system, amenable to widely available analysis techniques. The following analyses are presented (1) to help the reader develop a sound sense of engineering judgment, and (2) to provide a method of quick system evaluation should it be necessary.

5.4 Basic Approach

We will examine the proportional control of an integration and first-order system in cascade, resulting in a second-order position servo plant. We will derive the transfer function and place it into standard second-order form. Then the following analyses will be performed:

1. Response to a step-input command
2. Steady-state error for a step-input command
3. Response to a ramp-input command
4. Steady-state error for a ramp-input command
5. Response to a sinusoidal-input command

5.5 Transfer Function Development

From Chapter 3, the open-loop transfer function for a gear-reduced DC motor with an inertial load is

$$\frac{\theta_o}{e_i}(D) = nK_T/[L_aJ_{eq}D^3 + (B_{eq}L_a + R_aJ_{eq})D^2 + (R_aB_{eq} + n^2K_TK_e)D] \qquad (5.1)$$

Ignoring the armature inductance, this becomes

$$\frac{\theta_o}{e_i}(D) = nK_T/[(R_aJ_{eq})D^2 + (R_aB_{eq} + n^2K_TK_e)D] \qquad (5.2)$$

Adding the gain of a linear amplifier for proportional control, we have

$$\frac{\theta_o}{e_i}(D) = nK_AK_T/D[RJD + RB + n^2K_TK_e] \qquad (5.3)$$

where for simplicity we define the following variables:

$$R = R_a \qquad (5.4a)$$

$$J = J_{eq} = n^2J_M + J_L \qquad (5.4b)$$

$$B = B_{eq} = n^2B_M + B_L \qquad (5.4c)$$

The standard mathematical form for analysis of any second-order system is

$$\frac{O}{I}(D) = X/(D^2 + YD + X) \qquad (5.5)$$

where O is the output, I is the input, and

$$X \triangleq \omega_n^2 \qquad (5.5a)$$

$$Y \triangleq 2\zeta\omega_n \qquad (5.5b)$$

The quantity ω_n is the undamped natural frequency of the system, and the quantity ζ is the damping ratio. The second-order system is entirely defined by these two quantities. Our goal now is to place the closed-loop transfer function of the servo shown in Figure 5.1 into the form of (5.5). We know for a system with unity feedback,

$$\frac{\theta_o}{\theta_i}(D) = \frac{G}{1+G} \tag{5.6}$$

where G represents the open-loop transfer function of (5.3). The closed-loop transfer function is

$$\frac{\theta_o}{\theta_i}(D) = \frac{nK_AK_T}{nK_AK_T + D[RJD + RB + n^2K_TK_e]} \tag{5.7}$$

Placing this expression into the standard form of (5.5) results in

$$\omega_n = \sqrt{\frac{nK_TK_A}{RJ}} \tag{5.8}$$

$$\zeta = (n^2K_TK_e + RB)\frac{1}{[2(\sqrt{nK_TK_ARJ})]} \tag{5.9}$$

We note a trade-off in these expressions, in that the natural frequency is proportional to $\sqrt{K_A}$ and the damping ratio is proportional to $1/\sqrt{K_A}$. Thus, if the gain is increased to make the system faster (higher ω_n), it becomes more oscillatory (lower ζ).

5.6 Response to a Step-Input Command

The step input is very useful as a test signal [Kuo, 1982] due to its instantaneous jump in amplitude. It reveals a great deal about a system's quickness to respond. This is due to the fact that the step contains a wide band of frequencies in its spectrum and can be thought of as applying a wide range of sinusoidal signals to a system simultaneously. Given the system transfer function

$$\frac{\theta_o}{\theta_i}(D) = \frac{\omega_n^2}{D^2 + 2\zeta\omega_nD + \omega_n^2} \tag{5.10}$$

we can invoke Laplace transform notation and solve for the output as a function of s:

$$\theta_o(s) = \theta_i(s) \frac{\omega_n^2}{s^2 + 2\zeta\omega_n s + \omega_n^2} \tag{5.11}$$

The Laplace transform of the input step function is

$$\theta_i(s) = 1/s \tag{5.12}$$

The output as a function of s becomes

$$\theta_o(s) = \left(\frac{1}{s}\right) \frac{\omega_n^2}{s^2 + 2\zeta\omega_n s + \omega_n^2} \tag{5.13}$$

Taking the inverse Laplace transform of both sides, we get

$$\theta_o(t) = 1 + \frac{e^{-\zeta\omega_n t}}{\sqrt{1 - \zeta^2}} \sin(\omega_n t \sqrt{1 - \zeta^2} - \phi) \tag{5.14}$$

where

$$\phi = \tan^{-1}\left(\frac{\sqrt{1 - \zeta^2}}{-\zeta}\right) \tag{5.15}$$

The step response of the proportionally controlled position servo is plotted in Figure 5.2 for several values of ζ. How can we best interpret this figure? It is worthwhile to return to the s-plane and examine the roots of the characteristic equation of (5.10). These roots will change depending on the value of the amplifier gain K_A, since the variables ω_n and ζ both depend on K_A according to (5.8) and (5.9). Accordingly, the characteristic equation of (5.10) as a function of s is

$$s^2 + 2\zeta\omega_n s + \omega_n^2 = 0 \tag{5.16}$$

The roots of (5.16) are

$$s_1, s_2 = -\alpha \pm i\omega_d \tag{5.17}$$

where

$$\alpha = \zeta\omega_n \tag{5.18}$$

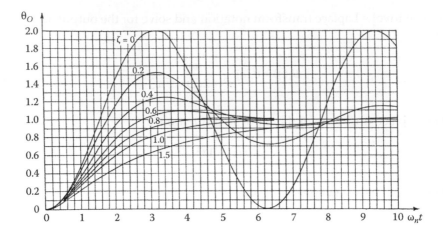

FIGURE 5.2
Step response of a proportionally controlled position servo.

and

$$\omega_d = \omega_n \sqrt{1 - \zeta^2} \tag{5.19}$$

Figure 5.3 shows the relationship between the location of the characteristic equation roots and the variables under discussion α, ζ, ω_n, and ω_d. The "damped" natural frequency ω_d is the frequency that can be actually observed in practice,

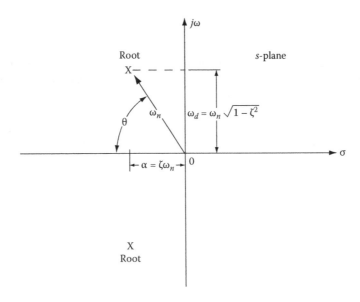

FIGURE 5.3
Relationship between characteristic equation roots and α, ζ, ω_n, and ω_d.

since all real systems have some damping. For the complex conjugate roots shown, ω_n is the radial distance from the origin to the roots, and concentric circles on the s-plane define loci of constant undamped natural frequencies. The damping factor α is the real part of the roots, and the damped natural frequency ω_d is the imaginary part. The damping ratio ζ is equal to the cosine of the angle between the radial line to the roots and the negative real axis, such that

$$\zeta = \cos\theta \qquad (5.20)$$

Radial lines at different angles on the s-plane define loci of constant damping ratios. The variable α is known as the damping factor, and it controls the rate of rise and decay of the time response. When the two roots are real and equal, the system is critically damped and $\zeta = 1$. Under this condition, the damping factor $\alpha = \omega_n$. Therefore, ζ is in fact a damping "ratio"—the ratio between the actual damping factor and the damping factor when the damping is critical. The effect of the characteristic root locations on the damping of our second-order servo can be seen in Figures 5.4 and 5.5. The natural frequency is held constant in Figure 5.4 while the damping ratio ζ is varied from $-\infty$ to $+\infty$. Kuo then classifies the system dynamics with respect to the value of ζ as shown in Table 5.1. Figure 5.5 shows typical step responses that correspond to the root locations given in the table.

From a control standpoint, the underdamped case ($0 < \zeta < 1$ in Figure 5.5) is of considerable interest [Doebelin, 1962]. The reason is that response speed specifications usually require that the error be reduced to a certain percentage of the step command within a certain time, while allowing some overshoot. Under these conditions, an underdamped system will be faster than a critically

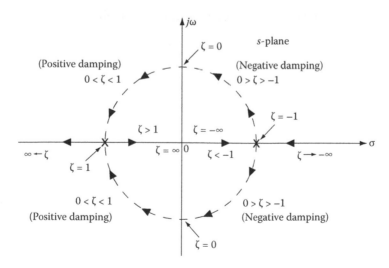

FIGURE 5.4
Locus of characteristic equation roots with ω_n constant and $-\infty < \zeta < +\infty$.

TABLE 5.1

Second-Order System Dynamics with Respect to Damping Ratio

Case Classifier	Damping Ratio	Roots of Characteristic Equation, s_1, s_2
Underdamped	$0 < \zeta < 1$	$-\zeta\omega_n \pm i\omega_n\sqrt{1-\zeta^2}$
Critically damped	$\zeta = 1$	$-\omega_n$
Overdamped	$\zeta > 1$	$-\zeta\omega_n \pm i\omega_n\sqrt{\zeta^2-1}$
Undamped	$\zeta = 0$	$\pm i\omega_n$
Negatively damped	$\zeta < 0$	$-\zeta\omega_n \pm i\omega_n\sqrt{1-\zeta^2}$

damped or overdamped system. Doebelin's guidance is that satisfactory values of ζ lie in the 0.4 to 0.8 range, corresponding to maximum overshoot percentages of 25 to 1.5 percent, according to the relation derived by [Kuo, 1982]:

$$\% \, maximum \, overshoot = 100e^{-\pi\zeta/\sqrt{1-\zeta^2}} \tag{5.21}$$

5.6.1 Steady-State Error Analysis for a Step Command

Figure 5.6 gives a more general block diagram of our position servo. The error signal is

$$\theta_e(s) = \theta_i(s) - \theta_b(s) \tag{5.22}$$

The feedback signal and output signal are, respectively,

$$\theta_b(s) = \theta_o(s)H(s) \tag{5.23}$$

$$\theta_o(s) = \theta_e(s)G(s) \tag{5.24}$$

By substitution, we have

$$\theta_b(s) = \theta_e(s)G(s)H(s) \tag{5.25}$$

Substituting (5.25) into (5.22), we get

$$\theta_e(s) = \theta_i(s) - \theta_e(s)G(s)H(s) \tag{5.26}$$

Solving for the error signal,

$$\theta_e(s) = \frac{\theta_i(s)}{1 + G(s)H(s)} \tag{5.27}$$

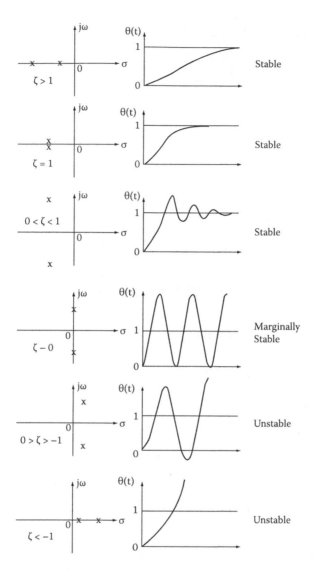

FIGURE 5.5
Typical step responses for root locations given in Table 5.1.

For unity gain potentiometer feedback,

$$\theta_e(s) = \frac{\theta_i(s)}{1 + G(s)} \tag{5.28}$$

Using the Laplace transform final-value theorem, we have

$$\theta_{e(ss)} = \lim_{t \to \infty} \theta_e(t) = \lim_{s \to 0} s\theta_e(s) \tag{5.29}$$

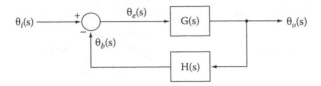

FIGURE 5.6
General block diagram of a position servo.

The Laplace transform for a step input of unity magnitude is

$$\theta_i(s) = 1/s \qquad (5.30)$$

Substituting this expression and (5.27) into (5.29), the steady-state error is

$$\theta_{e(ss)} = \lim_{s \to 0} \frac{1}{1 + G(s)H(s)} \qquad (5.31)$$

Defining the step error constant

$$K_{step} = \lim_{s \to 0} G(s)H(s) \qquad (5.32)$$

(5.31) becomes

$$\theta_{e(ss)} = \frac{1}{1 + K_{step}} \qquad (5.33)$$

The open-loop transfer function for our servo, assuming $H(s) = 1$ and using Laplace notation in (5.3), is

$$\frac{\theta_o}{e_i}(s) = (G)s = nK_A K_T / s[RJs + RB + n^2 K_T K_e] \qquad (5.34)$$

Substituting this expression into (5.32),

$$K_{step} = \lim_{s \to 0} G(s) = \infty \qquad (5.35)$$

Finally, the steady-state error for a unity-magnitude step-input command is

$$\theta_{e(ss)} = \frac{1}{1 + \infty} = 0 \qquad (5.36)$$

There is no steady-state error predicted for a position command input. However, Coulomb friction is present in all geared-down servos, which tends to make the output stick within a zone of uncertainty when the correction torque falls below the static friction level. A slight tendency of the output to oscillate when it comes close to agreement with the input can be used to remedy this situation. [Hayes and Horowitz, 2010] have also been successful in adding some electronic integration to the error signal to overcome this friction "dead-band."

5.7 Response to a Ramp-Input Command

Some DC position servos are called upon to track a command changing at a uniform rate. A position command changing at a uniform rate is a constant velocity signal. This type of command is usually used during long moves in the case of machine tools and large format plotters. A typical long move might consist of a positive acceleration section, a zero acceleration section, and a negative acceleration section, as shown in Figure 5.7. The velocity profile in the figure is known as a trapezoidal profile. The middle section of this profile, where the velocity is constant, is what we are discussing now.

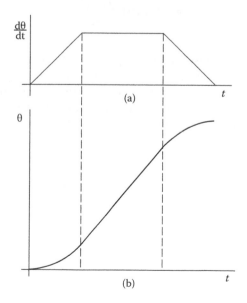

FIGURE 5.7
(a) Trapezoidal velocity profile and (b) corresponding position profile.

Repeating our system output as a function of s, we have

$$\theta_o(s) = \theta_i(s)\frac{\omega_n^2}{s^2 + 2\zeta\omega_n s + \omega_n^2} \qquad (5.11)$$

The Laplace transform of the input ramp function is

$$\theta_i(s) = 1/s^2 \qquad (5.37)$$

The output as a function of s becomes

$$\theta_o(s) = \left(\frac{1}{s^2}\right)\frac{\omega_n^2}{s^2 + 2\zeta\omega_n s + \omega_n^2} \qquad (5.38)$$

Taking the inverse Laplace transform of both sides, we get

$$\theta_o(t) = t - \frac{2\zeta}{\omega_n} + \frac{e^{-\zeta\omega_n t}}{\omega_n\sqrt{1-\zeta^2}}\sin\left(\omega_n t\sqrt{1-\zeta^2} - \phi\right) \qquad (5.39)$$

where

$$\phi = 2\tan^{-1}\left(\frac{\sqrt{1-\zeta^2}}{-\zeta}\right) \qquad (5.40)$$

A typical ramp response of the proportionally controlled position servo is plotted in Figure 5.8 for values of $\zeta < 1$. The underdamped case starts out with some oscillation and then tracks the input command with a certain level of steady-state error. It is important to note that this error is present only as long as the ramp command is given. When the command ceases, the

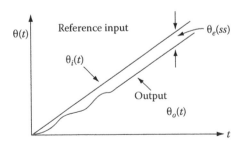

FIGURE 5.8
Ramp response of proportionally controlled position servo.

position error will be integrated to zero. Suppose a ramp command is given for an extended period in a rotary radar antenna tracking system. If position is the controlled variable, the ramp command is asking the system to rotate at a constant velocity. The actual velocity might equal the commanded velocity; however, the actual position will lag behind the command. In this case, we have a tracking error, which must be accounted for in radar systems seeking a target in real time. We now proceed to calculate this error.

5.7.1 Steady-State Error Analysis for a Ramp Command

For a ramp input of unity magnitude,

$$\theta_i(s) = 1/s^2 \tag{5.41}$$

Substituting this expression and (5.27) into (5.29), the steady-state error is

$$\theta_{e(ss)} = \lim_{s \to 0} \frac{1}{s[1 + G(s)H(s)]} \tag{5.42}$$

For gains greater than unity, we can approximate (5.42) as

$$\theta_{e(ss)} = \lim_{s \to 0} \frac{1}{sG(s)H(s)} \tag{5.43}$$

Defining the ramp error constant for unity feedback gain

$$K_{ramp} = \lim_{s \to 0} sG(s) \tag{5.44}$$

(5.43) becomes

$$\theta_{e(ss)} = \frac{1}{k_{ramp}} \tag{5.45}$$

The open-loop transfer function for our servo has not changed, that is,

$$G(s) = nK_A K_T / s[RJs + RB + n^2 K_T K_e] \tag{5.34}$$

Substituting this expression into (5.44),

$$K_{ramp} = \lim_{s \to 0} sG(s) = nK_A K_T / [RB + n^2 K_T K_e] \tag{5.46}$$

Finally, the steady-state error for a unity-magnitude ramp-input command is

$$\theta_{e(ss)} = \frac{1}{K_{ramp}} = [RB + n^2 K_T K_e]/nK_A K_T \qquad (5.47)$$

This error is inversely proportional to the amplifier gain, so that increasing the gain decreases the tracking error. This comes at the expense of a more oscillatory step response.

5.8 Response to a Sinusoidal-Input Command

Although DC servos are not usually called upon to track fast sinusoidal signals due to their slow response times, a prediction of the system's frequency response is still useful. The work of [Black, 1934] and [Bode, 1945], originally in the telephone communication field, gradually made its way into control system theory because of the common element of negative feedback. It has allowed control engineers working with inherently slow systems to communicate with engineers involved in higher-speed circuit design including communication circuits. We will also demonstrate later how the frequency response method can serve as a check on the stability of op-amps used in real control circuits. Now, we concentrate on the frequency response of our proportionally controlled servo for a sinusoidal-input signal swept over a range of frequencies from low to high. Generally the output signal will have a change in amplitude and a phase shift when compared to the input, and some engineers tend to think of servos as electromechanical "filters." When looking at the Bode plots (amplitude and phase) of a DC servo, they see no real distinction between it and an electrical filter.

From (5.10), using Laplace notation we have

$$\frac{\theta_o}{\theta_i}(s) = \frac{\omega_n^2}{s^2 + 2\zeta\omega_n s + \omega_n^2} \qquad (5.48)$$

For pure sinusoidal excitation, $s = i\omega$ and

$$\frac{\theta_o}{\theta_i}(i\omega) = \frac{\omega_n^2}{(i\omega)^2 + 2\zeta\omega_n i\omega + \omega_n^2} \qquad (5.49)$$

To make a plot of this transfer function more universal, the frequency ratio ω/ω_n is used to normalize the frequency axis, that is,

$$\frac{\theta_o}{\theta_i}(i\omega) = \frac{1}{[1 - (\omega/\omega_n)]^2 + i\left(2\zeta\frac{\omega}{\omega_n}\right)} \qquad (5.50)$$

The magnitude of (5.50) is the amplitude ratio of output to input. If we simplify (5.50) with

$$u = \omega / \omega_n \tag{5.51}$$

the amplitude ratio is

$$\frac{A_{\theta o}}{A_{\theta i}}(u) = \frac{1}{[(1-u^2)^2 + (2\zeta u)^2]^{1/2}} \tag{5.52}$$

The phase angle by which the output leads the input is given by

$$\phi(u) = \tan^{-1}\left[-\frac{2\zeta u}{1-u^2}\right] \tag{5.53}$$

The negative sign in the argument of (5.53) is the result of the complex variable i being in the denominator of (5.50). Normalized amplitude ratio and phase-shift curves for a second-order system are given in Figures 5.9 and 5.10, respectively.

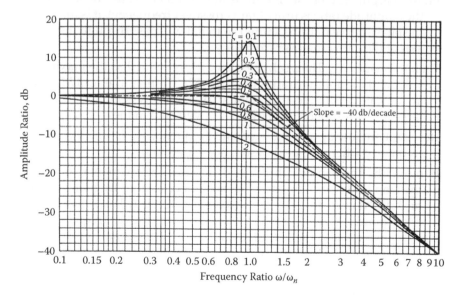

FIGURE 5.9
Normalized amplitude-ratio curves for a second-order system.

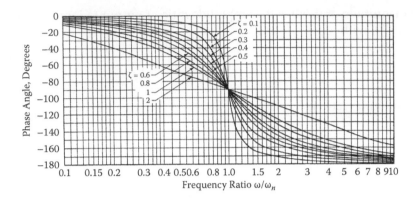

FIGURE 5.10
Normalized phase-shift curves for a second-order system.

The typical gain and phase characteristics of a feedback control system are shown in Figure 5.11. Following Kuo, we now define some frequency response characteristics often used in practice, with reference to Bode plots:

Gain margin. The gain margin is measured at the frequency where the phase lag between input and output is 180° (phase crossover frequency). If the loop gain of the system is increased by more than this amount at the phase crossover frequency, the system becomes unstable.

Phase margin. The phase margin is measured at the frequency where the amplitude ratio is unity (gain crossover frequency). It is the angle the phase curve must be shifted so that it will pass through the –180° axis at the gain crossover frequency.

Peak resonance M_p. The peak resonance M_p is defined as the maximum value of the amplitude ratio given in (5.52). The magnitude of M_p gives an indication of the relative stability of a servo system. Specifically, a large M_p corresponds to a large peak overshoot in the step response.

Resonant frequency ω_p. The resonant frequency ω_p is defined as the frequency where the peak resonance M_p occurs.

Bandwidth. The bandwidth BW is defined as the frequency at which the amplitude ratio drops to 70.7 percent of its zero-frequency level, or 3 dB "down" from its zero-frequency level. A large bandwidth corresponds to a faster system rise time, since higher-frequency signals are passed to the output. If the bandwidth is small, only lower-frequency signals are passed, making the output have a relatively slow transient response.

It can be shown [Kuo, 1982] that for a second-order system, the quantities M_p, ω_p, and BW are all uniquely related to the damping ratio ζ and the undamped natural frequency ω_n.

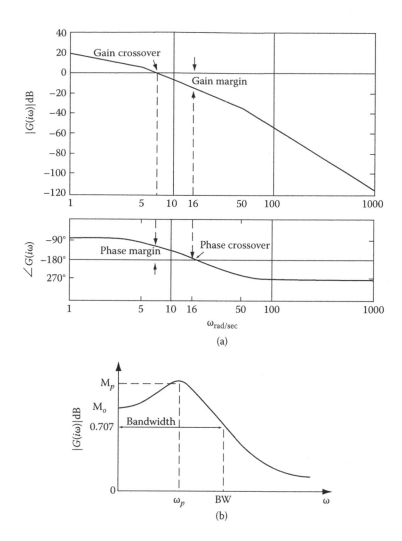

FIGURE 5.11
(a) Typical gain and phase characteristics of a feedback control system and (b) gain characteristic details.

References

Black, H., "Stabilized feedback amplifiers," *Bell Sys. Tech. J.*, v. 13, 1934.
Bode, H., *Network Analysis and Feedback Amplifier Design*, Van Nostrand, 1945.
Doebelin, E., *Dynamic Analysis and Feedback Control*, McGraw-Hill, 1962.
Hayes, T., and P. Horowitz, *Learning the Art of Electronics*, Cambridge, 2010.
Kuo, B., *Automatic Control Systems*, Prentice-Hall, 1982.

6

Compensation of a Continuous-Time DC Servo

6.1 Introduction

We know from the study of a second-order servo that critical damping is achieved when the roots of the system characteristic equation are real and equal. Because the critically damped case is the dividing line between an underdamped system and an overdamped system, there is only one critically damped system. Critically damped performance is desirable because it represents the fastest system that will not overshoot. However, getting the two roots to equalize by manipulating the system variables is extremely difficult to achieve in practice. Once manipulated, variables may drift over time, affecting the transient response. There are an infinite number of underdamped and overdamped systems, and within these systems, many combinations of system parameters.

Although critical damping may not be a realistic goal, it is possible that proportional control can give adequate results. Consider the shaft repeater introduced earlier in the book, using proportional control only. Without adequate gain, the system is overdamped and suffers from steady-state error due to stiction. In this condition it might settle in 2.0 sec. With increased gain, the system is underdamped, and the steady-state error is reduced because of the oscillation about the target position. In this new condition the system might take 1.0 sec to settle. However, what if we wanted (or needed) better transient performance, say, a slightly underdamped response with settling time no greater than 0.5 sec? We essentially have two choices: (1) change the motor to one with a higher torque-to-inertia ratio, or (2) augment the controller by compensation.

In many cases, the control engineer chooses (or is given) the "plant" at the start of the design process. For example, a DC motor may have been selected for a servo application because it was the only one available from a vendor at the time of its choosing. This means that most system parameters become immediately fixed. Some of these parameters may be known; others may be

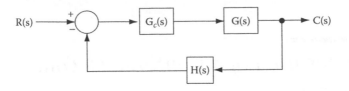

FIGURE 6.1
A compensated controller in cascade with a plant.

unknown. Compensating a proportional controller with electronic circuitry is a powerful tool we can use to tailor the transient response to our needs, without changing the system variables. We will now explore the different ways of compensating a proportional controller, such that critically damped performance is the ideal goal, but a slightly underdamped response is a more realistic result. We will also demonstrate a compensation scheme with adjustable properties, to accommodate unknown values of the plant variables. The generalized form of a feedback control system comprising a compensated controller $G_c(s)$ in cascade with a plant $G(s)$ is shown in Figure 6.1.

6.2 Compensation Using Derivative Control

Consider the control system error-time curve of Figure 6.2. As [Doebelin, 1962] suggests, if proportional control were being used, the controller would exert the same corrective effort at point B as at point A, even though the error is increasing at A and decreasing at B. A human operator might note the trend of the curve and increase the correction as the error gets larger at A, while decreasing the correction as the error gets smaller at B. A human noting the trend of the error-time curve can be approximated by a machine sensing the slope, or derivative, of the curve. This can be done by differentiating the error signal with respect to time and adding it to a proportional term. As shown in the figure, a two-part controller of this type can improve settling time as a result of a transient input to the system. This so-called P-D controller can be realized in block diagram form as shown in Figure 6.3(a). This particular case shows two transfer blocks for the P and D sections arranged in parallel, although other arrangements are possible. Derivative control must be implemented with a proportional term, since the slope of the error-time curve can be zero when the error itself is nonzero, thus offering no corrective action by itself. The P-D controller transfer function is

$$G_c(s) = K_d s + K_p = K_d \left(s + \frac{K_p}{K_d} \right) \tag{6.1}$$

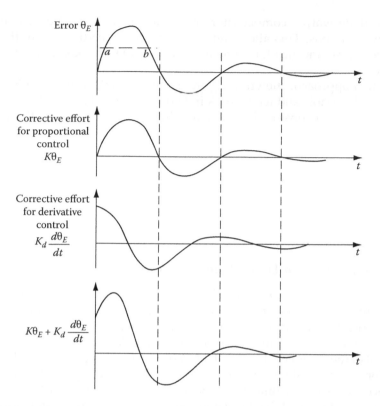

FIGURE 6.2
Action of proportional, derivative, and P-D control.

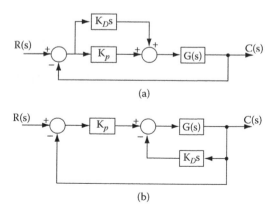

FIGURE 6.3
Block diagrams of a P-D controller (a) derivative of error and (b) derivative of controlled variable.

An ideal derivative compensator [Nise, 2004] has a single zero on the negative real axis. This allows an improvement (speeding up) of the transient response and has the same effect as adding viscous damping to the system.

Another approach, shown in Figure 6.3(b), is to take the derivative of the controlled variable and feed it back in addition to feedback of the controlled variable itself. In positioning systems, this had been done in the past by using a DC tachometer. More recently, this approach has been replaced by counting the pulses of an incremental encoder over a specific length of time, computing a velocity, and repeating the process continuously until the target position has been attained.

6.3 Compensation Using Integral Control

Improving steady-state error can be accomplished by adding an accumulating or integrating element to a proportional controller. A human operator might note the error-time curve not closing fast enough and increase corrective action over time until the error is reduced. As shown in Figure 6.4, the integral of the error is fed to the plant in parallel with a proportional term. In position control systems, where no steady-state error is predicted in theory, practice shows it will almost always exist due to Coulomb friction, which is very tedious to model due to its nonlinearity. The integral term can be used to overcome Coulomb friction effects if proper precautions are used. The P-I controller transfer function is

$$G_c(s) = K_p + \frac{K_i}{s} = \frac{K_p\left(s + \frac{K_i}{K_p}\right)}{s} \tag{6.2}$$

An ideal integral compensator [Nise, 2004] has a pole at the origin and zero close to the pole. This allows improvement in the steady-state error without appreciably affecting the transient response.

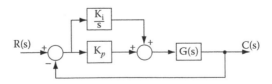

FIGURE 6.4
P-I controller block diagram.

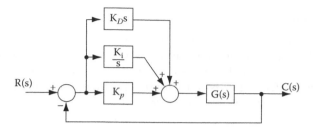

FIGURE 6.5
P-I-D controller block diagram.

6.4 Compensation Using Derivative and Integral Control

Reductions in both transient response time and steady-state error can be accomplished by combining the two aforementioned compensation schemes. A derivative term and an integral term can both be placed in parallel with a proportional term, as shown in Figure 6.5. The P-I-D controller transfer function is

$$G_c(s) = K_p + \frac{K_i}{s} + K_d s = \frac{K_d s^2 + K_p s + K_i}{s} \qquad (6.3)$$

6.5 Tools for Predicting Performance

We have three main tools at our disposal to predict the performance of a given DC servo design. These are root locus, Bode, and transient response plots. These are all graphical in nature, but how the plots are created in practice can vary. Theoretical plots are all available by invoking "sisotool" at the MATLAB command line. This stands for "single-input single-output tool" and is not only a graphing tool but also an interactive graphical user interface for control system design.

6.5.1 Root Locus

Root locus is a picture of stability in the s-plane. It is a predictive tool used to assess the stability of a chain of open-loop subsystems, assuming that a 100 percent feedback loop is closed around them. It is the solution of the characteristic equation $1 + KG = 0$ for all positive gain values zero to infinity. If the locus of roots crosses into the right half of the s-plane for a given gain, it is theoretically unstable at that gain.

6.5.2 Bode Plot

This is a picture of the frequency responses of both amplitude ratio and phase shift between output and input for a chain of open-loop subsystems. It is a predictive tool used to assess system stability over a range of frequencies. Although a DC servo is not usually commanded to track a sinusoidal input, other more abrupt inputs are composed of a superposition of many sine wave inputs. A gain margin and phase margin must be evident on the plots to ensure stability. These margins have a one-to-one correspondence with the transient response of a given system once the loop is closed.

6.5.3 Transient Response

This is a plot of the closed-loop system response to an abrupt input such as a ramp, step, or impulse. The most common input for assessing system performance is the leading or trailing edge of a pulse, known as a step input. Because it is made up of many sinusoidal waves, a single step can be used to assess response over a wide frequency range with relatively little labor. It is also the easiest test to be done in the laboratory; all that is needed is a function generator and an oscilloscope or strip chart recorder. The step response can also be calculated theoretically but is equally useful as a design tool when applied directly to the system and the results observed.

6.6 Overall Compensation Strategy

We wish to compensate the follow-up shaft repeater of Chapter 2 by replacing the network originally used as a placeholder in our design. Compensation can be done theoretically, by trial and error, or through a combination of both. The geared motor selected for the application consumes less than 5 W. One problem with motors this size is getting good data from manufacturers on inertia, damping, and torque constant. In this case, we propose to approximate the transfer function of the motor using the past experience of others. Because we are using op-amps as feedback amplifiers, we will take a look at how they are used within our control circuitry. We will then compensate the system by theoretical prediction and follow this by synthesizing a P-D controller in electronic circuitry. This controller has decoupled parameters [Tietze and Schenk, 1991] making adjustment of K_p and K_d non-interactive. Others [Hayes and Horowitz, 2010] have implemented integral control with success to solve the stiction problem in position control servos, although it has a destabilizing influence. To ensure stability, integral control will not be implemented, because natural integration from velocity to position is already present in the plant dynamics.

6.7 Op-Amps and Control Systems

A noninverting op-amp circuit with finite gain is an example of a feedback control system. We will not treat the inverting case explicitly here; the interested reader is referred to [Fredricksen, 1988]. As shown in Figure 6.6, the op-amp itself can be represented by the summing junction and gain block A (often called G in control work). The passive feedback network made up of R_1 and R_2 can likewise be represented by a feedback block β (often called H). The feedback factor β is the fraction of the output which is fed back to the input. It can range from 0 (no feedback) to 1 (100 percent feedback). The three cases shown in Figure 6.7 illustrate this point. We assume that these circuits have reached their steady-state values and all transients have died away. Figure 6.7(a) has no feedback, $\beta = 0$, and

$$V_{out} = AV_{in} \tag{6.4}$$

Because of the large value of open-loop gain A in the op-amp, the output will saturate at its positive power supply rail. This is an uncontrolled amplifier, and not very useful to us. The other extreme is shown in Figure 6.7(b). Here we have 100 percent feedback, $\beta = 1$, and

$$V_{out} = V_{in} \tag{6.5}$$

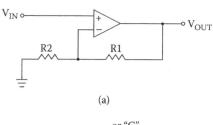

(a)

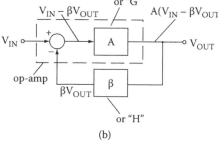

(b)

FIGURE 6.6
Noninverting op-amp (a) circuit analysis view and (b) control system view.

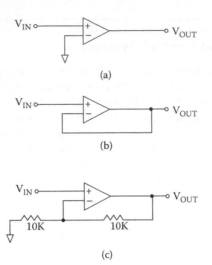

FIGURE 6.7
Noninverting op-amp with (a) $\beta = 0$, (b) $\beta = 1$, and (c) $\beta = 0.5$.

In this case the output "follows" the input with as little deviation as possible. This is the simplest form of a controller we have seen. As an intermediate case, consider the 50 percent feedback factor of Figure 6.7(c). The feedback factor β is simply a voltage divider situation where $R_1 = R_2$ and

$$\beta = \frac{R_2}{R_1 + R_2} = 0.5 \tag{6.6}$$

As the feedback factor increases, the gain of the circuit decreases. The gain is made finite and predictable with feedback using passive components. It is a feedback amplifier. This is similar to the servo we are trying to compensate, in that our feedback is made up of passive resistors and potentiometers. For the 50 percent feedback case, let us examine the closed-loop gain from two perspectives. The control system view is

$$A_{cl} = \frac{V_{out}}{V_{in}} = \frac{A}{1 + A\beta} = \frac{G}{1 + GH} \tag{6.7}$$

For $A \gg 1$, we get the approximation

$$A_{cl} \cong \frac{1}{\beta} = 2 \tag{6.8}$$

In the circuit analysis view, we apply KCL to the inverting input of the op-amp. Two assumptions are important: (1) no current enters or leaves the inverting input, and (2) the difference between the two op-amp inputs is driven to zero by the feedback. We have

$$\frac{V_{out} - V_{in}}{R_1} = \frac{V_{in} - 0}{R_2} \tag{6.9}$$

Rearranging this for $R_1 = R_2$ we get

$$\frac{V_{out}}{V_{in}} = A_{cl} = 2 \tag{6.10}$$

In the control system view, we assumed that the open-loop gain was close to infinite and the summing (difference) junction was ideal. In the circuit analysis view, we assumed the deviation between the two op-amp inputs was zero and no current entered the inverting input. In a real circuit, for any feedback factor including unity there is a "gain error." For a closed-loop gain of 2 ($\beta = 0.5$), we can compute a finite difference between the two inputs of the op-amp for a typical open-loop gain of 10^5. Assuming now V_{in-} is an unknown quantity,

$$A(V_{in+} - V_{in-}) = 2V_{in+} \tag{6.11}$$

With $A = 10^5$ and $V_{in+} = 1$ V, solving for V_{in-} we get

$$V_{in-} = \frac{(A-2)V_{in+}}{A} = 0.99998 \ V \tag{6.12}$$

The point here is that the difference is not driven completely to zero. The actual output is dictated by the feedback network, and

$$V_{out} = 2V_{in-} = 1.99996 \ V \tag{6.13}$$

The gain error is

$$\left[\left(\frac{V_{out,ideal}}{V_{out,actual}} \right) - 1 \right] (100\%) = .002\% \tag{6.14}$$

This computation (6.14) is the steady-state error of a typical op-amp circuit with 50 percent feedback.

6.7.1 A Control System within a Control System

Returning to the control system view, for a feedback amplifier we have for the closed-loop gain

$$A_{cl} = \frac{V_{out}}{V_{in}} = \frac{A}{1 + A\beta} \qquad (6.15)$$

For our DC servo to remain stable, all the feedback amplifiers within it must remain stable. To cause one of the amplifiers to become unstable, the closed-loop gain must be made to increase without bound at some frequency. This occurs when the denominator of (6.15) approaches zero, such that in the limit

$$1 + A\beta = 0 \qquad (6.16)$$

Rearranging (6.16), we define the "loop gain" $A\beta$ and get

$$A\beta = -1 \qquad (6.17)$$

as a condition for instability. This is really two separate conditions involving the loop gain combined into one. In general, the open-loop gain A has a magnitude and a phase shift associated with it. For example, the op-amp we are using in the shaft repeater design is the TL082 dual IC from Texas Instruments. Its magnitude and phase plots are shown together in Figure 6.8. Like most op-amps, it behaves as a low-pass filter with a very

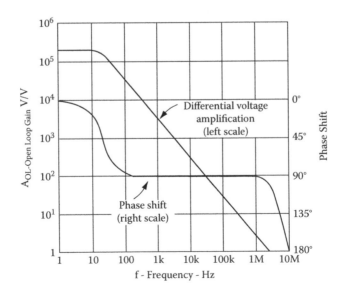

FIGURE 6.8
Open-loop gain and phase shift vs. frequency for the TL082 operational amplifier.

wide bandwidth. For instability, both of the following loop gain relations must be satisfied simultaneously:

$$A\beta = 1 \tag{6.18}$$

$$\angle A\beta = 180° \tag{6.19}$$

Equation (6.18) is known as the magnitude condition, and (6.19) is known as the phase condition. Furthermore, having all amplifiers within the servo stable, it is desirable that each one contribute as little phase shift as possible to the overall servo loop.

To gain some insight into the stability problem, we first examine the amplitude ratio and phase-shift Bode plots in Figure 6.9 for the ideal case of a feedback amplifier. For this discussion, it is assumed that the feedback network is composed of resistors only. The open-loop gain of an op-amp (or any amplifier we choose) usually follows a curve similar to that shown in Figure 6.9(a). Not so obvious is the notion that the area within the plot can be divided into the closed-loop gain and the loop gain $A\beta$. From the logarithmic nature of

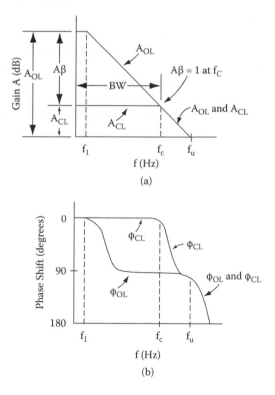

(a)

(b)

FIGURE 6.9
Open-loop and closed-loop op-amp Bode plots for (a) gain and (b) phase shift.

the plot, there can be no greater gain in the system than the open-loop gain of the amplifier, and

$$logA_{ol} = logA_{ol}\beta + logA_{cl} \tag{6.20}$$

Rearranging (6.20) we have

$$logA_{ol}\beta = logA_{ol} - logA_{cl} \tag{6.21}$$

Taking the antilog of both sides, (6.21) is equivalent to the expression

$$A_{ol}\beta = \frac{A_{ol}}{A_{cl}} \tag{6.22}$$

In words, the loop gain is equal to the open-loop gain divided by the closed-loop gain. The closed-loop gain and the open-loop gain are equal at the crossover frequency f_c, so by definition the loop gain is unity at this frequency. The span from DC to f_c is the bandwidth of the amplifier. The closed-loop amplifier response is flat up to f_c and then rolls off at 20 dB per decade to the unity-gain frequency f_u. Turning now to Figure 6.9(b), we can intuitively reason that any phase shift is the result of a roll-off in the amplitude response. For the open-loop op-amp, the phase begins to shift at Fredricksen's "first pole," f_1. When the loop is closed, the gain is reduced and the response is flat up to f_c, which causes the phase shift to be zero up to f_c. As long as the closed-loop amplifier is operated well below f_c, it will remain stable and its phase shift will be negligible.

6.7.2 Going around the Servo Loop

Because we are using a series of op-amps with a current booster to drive our DC servo, they are intimately related within our system. A block diagram is shown in Figure 6.10, with narrative descriptions of each block. It is imperative that the loop gain of the servo be less than unity at the frequency where the total phase shift around the loop reaches 180°. For example, the blocks F, H, and S in Figure 6.10 are usually assumed to be stable themselves and add no phase shift to the overall loop gain of the servo. Let us examine these to evaluate their individual contributions to the servo's stability. F and H are passive networks with low source impedance. However, to prevent any loading, each is connected to an op-amp follower whose input impedance is extremely high. Because noise can easily be picked up at areas of high impedance, we have deliberately rolled off the frequency response of the input networks to attenuate any high-frequency noise. For instance, each of the input network potentiometers can act like an antenna. Placing capacitors

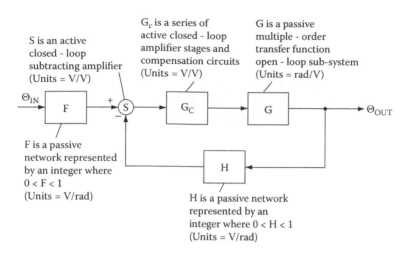

FIGURE 6.10
Control loop block descriptions for a DC servo.

as close as possible to the input networks, rolling the networks off between 10 Hz and 100 Hz, mitigates this potential problem.

The closed-loop subtracting amplifier *S* we assume will be stable, since the circuit is unity-gain and the amplifier used is unity-gain stable. Its transfer function can be derived using the method of superposition [Franco, 1988]. The voltage followers buffering each input of the subtracting amplifier prevent loading of the input networks. Returning to Figure 6.9, for the follower

$$A_{cl} = 1 \tag{6.23}$$

$$A\beta = A_{ol} \tag{6.24}$$

The loop gain of the follower is as large as it can be, and the bandwidth the widest it can be. However, its breakpoint is the unity-gain frequency of the op-amp. This breakpoint corresponds to as large a phase shift as possible. For the follower to be stable across its entire bandwidth, the op-amp must be unity-gain stable. In the case of the DC servo, high-frequency input to the followers can be injected by noise pickup from the outside world if it is not attenuated in advance. Both the followers and subtracting amplifier, if operated well below f_u, will add negligible phase shift to the servo loop.

The logarithmic amplitude ratio plots of the compensated amplifier G_c and a motor *G* in series combination can simply be added together to assess overall stability. The plots of Figure 6.11 show how each subsystem is represented as it is used in our DC servo design. For example, using proportional control alone the G_c block is a gain block. Figure 6.11(a) shows a typical op-amp closed-loop gain plot for a gain of 100 (40 dB). Its closed-loop bandwidth is

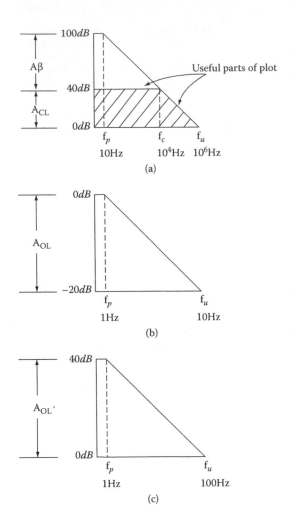

FIGURE 6.11
Subsystem Bode plot representation of (a) closed-loop amplifier, (b) open-loop DC servo, and (c) the sum of the two (note scale changes).

10^4 Hz, and it contributes zero phase shift below this frequency. Figure 6.11(b) shows a somewhat simplified DC motor/load open-loop gain plot, representing the block G. Its amplitude spans 20 dB, and its closed-loop bandwidth is 10 Hz. This plot begins at 0 dB because a motor and load does not generate energy unless it is amplified. Another way to visualize this is to look at the open-loop DC motor/load transfer function, which is made up entirely of passive elements, such as resistance, inductance, inertia, and damping. It has unity gain at low frequencies and behaves as a low-pass filter as the frequency is increased. Adding the plots of Figures 6.11(a) and (b) yields the plot of Figure 6.11(c). Its magnitude spans 40 dB at DC and rolls off at the same rate as the motor alone up to 100 Hz. In the case of proportional control only, the phase shift is defined solely by the motor/load block G.

6.8 Compensation by Theoretical Prediction

The transfer function of our third-order position servo can be expressed in factored form as

$$G(s) = \frac{1}{s(\tau_e s + 1)(\tau_m s + 1)} \tag{6.25}$$

Converting to MATLAB's zero-pole-gain form,

$$G(s) = \frac{1/\tau_e \tau_m}{s\left(s + \frac{1}{\tau_e}\right)\left(s + \frac{1}{\tau_m}\right)} \tag{6.26}$$

Using the experience of others [Doebelin, 1962], we now estimate the electrical and mechanical time constants of a typical instrument servo motor as

$$\tau_e = \frac{L}{R} = 0.01 \ sec \tag{6.27a}$$

$$\tau_m = \frac{J}{B} = 0.1 \ sec \tag{6.27b}$$

Substituting these values into (6.26), we get for the servo plant approximation

$$G(s) = \frac{1000}{s(s + 10)(s + 100)} \tag{6.28}$$

Figure 6.12 shows the open-loop root locus and Bode plots for this plant at a DC gain of unity. Its gain margin is 40.8 dB and phase margin 83.7°. The factor of 1000 in the numerator of (6.28) is the result of a change in mathematical form, not something physical. We will therefore take its gain reference to be 0 dB at 1 rad/sec (0.16 Hz), because we know the plant can only attenuate signals, not amplify them. A gain of 12.5 and 100 percent feedback results in the closed-loop step response shown in Figure 6.13. With 30 percent overshoot, the settling time is approximately 1 sec. Our goal now is to compensate this servo plant to improve the settling time by a factor of at least 5 with the same overshoot percentage, using a P-D controller scheme. This is done by adding a negative real zero to the compensator transfer function. After some trial and error within the MATLAB sisotool GUI, a suitable compensator with gain turns out to be

$$G_c(s) = 200(1 + 0.05s) \tag{6.29}$$

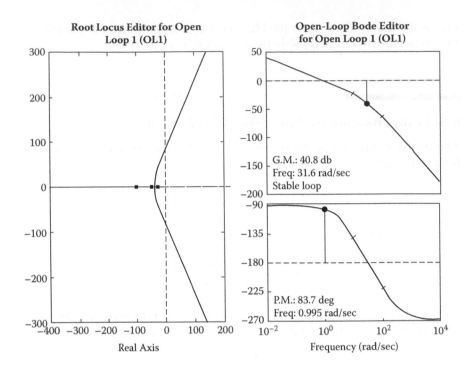

FIGURE 6.12
Open-loop root locus and Bode plots for DC servo, $K = 1$.

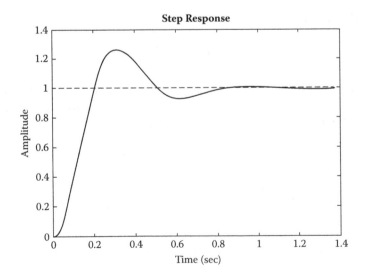

FIGURE 6.13
Closed-loop step response of DC servo, $K = 12.5$.

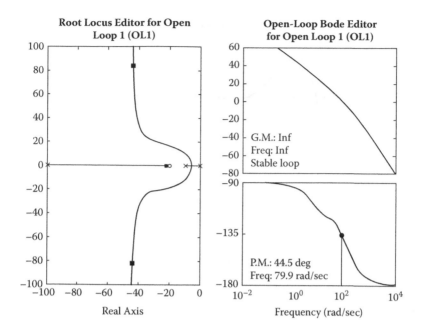

FIGURE 6.14
Open-loop root locus and Bode plots for compensated DC servo system.

The root locus and Bode plots of the compensated system are given in Figure 6.14. The real zero, denoted by the "o" symbol, is located at −20 rad/sec (3.18 Hz). The gain margin is improved to infinity, because the phase shift never reaches 180°. The phase margin is lower than the plant alone, but it is pushed out to almost 80 rad/s. We know this is beyond the bandwidth of our DC servo, and at lower excitation frequencies the phase margin approaches 90°. Notice the new trajectory of the root locus originating between the two poles closest to the origin. It has been pulled away from the positive half-plane, ensuring stability for higher gains. As shown in Figure 6.15, for the same 30 percent overshoot, the system settles in about 0.15 sec. This is a 6-to-1 improvement, certainly acceptable for our purposes here. The question now is how to make a real amplifier and compensator fit the prescription of (6.29).

6.8.1 Synthesizing a P-D Controller

We know the transfer function of a P-D controller is of the general form

$$G_c(s) = K_d s + K_p \qquad (6.30)$$

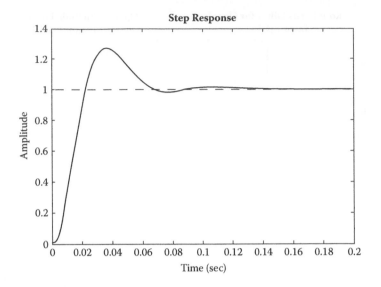

FIGURE 6.15
Closed-loop step response for compensated DC servo system.

A convenient method for realizing this controller is given in the block diagram of Figure 6.16. In the diagram, we propose to have two separate proportional gain blocks, K_{p1} and K_{p2}. Let the following values be assigned to the blocks:

$$K_d = 0.05$$

$$K_{p1} = 1$$

$$K_{p2} = 200$$

If $K_d s$ is an inverting differentiator, K_{p1} needs to be a unity-gain inverter. Then, if the total DC gain applied to the plant is –200, the controller will follow (6.29) exactly.

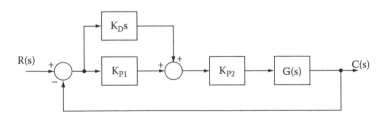

FIGURE 6.16
Revised block diagram for P-D controller.

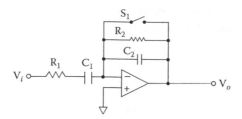

FIGURE 6.17
Practical differentiator, per National Semiconductor AN-20.

We can synthesize an inverting differentiator for our compensator by using the "practical differentiator" circuit of [National Semiconductor, 1980]. This circuit can be designed to behave as a differentiator up to a few kilohertz and then have a roll-off like an integrator beyond the breakpoint. This gives the circuit noise immunity at high frequencies. Even though our servo will not respond to it, high-frequency amplified noise can result in oscillation if not compensated. This in turn could render our amplifiers unable to do the low-frequency work we need them to do. The practical circuit is shown in Figure 6.17, with a switch to render the circuit inactive when closed. With the switch open, we have for the transfer function

$$\frac{V_o}{V_i}(s) = -\frac{R_2 C_1 s}{(R_2 C_2 s + 1)(R_1 C_1 s + 1)} \tag{6.31}$$

For our servo, the numerator of (6.31) represents part of the real zero we need for compensation (the other part is a unity gain inverter in parallel). The denominator represents the higher-frequency roll-off terms required to make the circuit practical. Ignoring the circuit inversions for the moment, we have chosen

$$R_2 C_1 = 0.05 \tag{6.32}$$

Choosing a common value for C_1 gives

$$C_1 = 0.1 \, \mu F$$

$$R_2 = 500 \, K\Omega \text{ nominal}$$

To adjust the gain of the differentiator, for R_2 we select a 1 MΩ potentiometer. Tietze and Schenk provide another method for choosing the real zero, an oscillation test of the uncompensated servo on the bench. The gain is increased until the output shows just slightly damped oscillation. The oscillation frequency gives a good starting point for choosing the real zero. We then choose a convenient roll-off point for the higher-frequency noise immunity, 10^4 rad/sec (1592 Hz). This exceeds the servo bandwidth enough so that the integrating action will only attenuate out-of-band noise. We have

$$R_1 C_1 = R_2 C_2 = 10^{-4} \tag{6.33}$$

This gives the values

$$R_1 = 1\ K\Omega$$

$$C_2 = 200\ pF\ \text{nominal}$$

We choose a commonly available value of 220 pF for C_2. The Bode plot for this controller is shown in Figure 6.18. We might now ask if the "practical differentiator" circuit introduces any inaccuracies in our controller. To answer this question, let us look at a more exact layout of the controller schematic in Figure 6.19. As long as the resistors feeding K_{p2} are equal and K_{p1} is unity gain, the sum of K_d and K_{p1} is

$$1 + G_d = \frac{R_2 C_1 s + (R_2 C_2 s + 1)(R_1 C_1 s + 1)}{(R_2 C_2 s + 1)(R_1 C_1 s + 1)} \tag{6.34}$$

where

$$G_{d,ideal} = K_d s = 0.5s \tag{6.35}$$

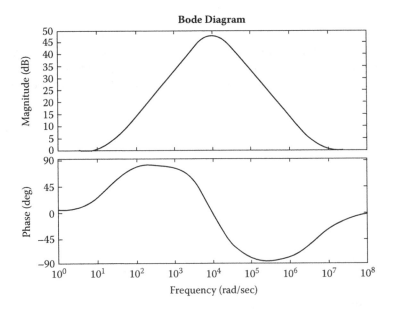

FIGURE 6.18
Bode plot for P-D controller using practical differentiator.

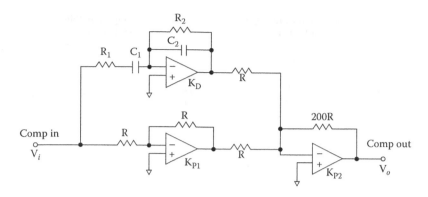

FIGURE 6.19
"Practical" circuit layout for P-D controller.

Substituting the component values we selected into (6.34), we get

$$1+G_d = \frac{10^{-8}s^2 + .05s + 1}{10^{-8}s^2 + .0002s + 1} \qquad (6.36)$$

The following code was used within MATLAB to arrive at a more exact solution:

```
%Exact compensator 2R's and 2C's
numcomp=[1e-8,.05,1];
dencomp=[1e-8,2e-4,1];
comp_tf=tf(numcomp,dencomp)
sys_dc_zpk=zpk([],[-100 -10 0],1000)
comp_zpk=zpk(comp _ tf)
sys_dc_comp=series(comp_zpk,sys_dc_zpk)
sisotool(sys_dc_comp)
%add DC gain of 200 within sisotool
```

After the DC gain was added within sisotool, the open-loop root locus and Bode plots were plotted with unity feedback; these are shown in Figure 6.20. The gain margin is 33.1 dB at 669 rad/sec. The phase margin and root loci remain unchanged. The closed-loop step response is shown in Figure 6.21. It is virtually identical to Figure 6.15, which proves that making the differentiator "practical" has no effect on the compensated servo performance.

6.8.2 Schematic Changes

The original compensator and gain stages U3A and U3B of Figure 2.4 have been replaced on a second schematic page. The final schematic for our shaft repeater servo is given in Figures 6.22 and 6.23. The gain of U3B can be varied

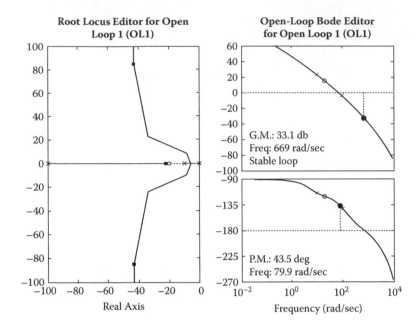

FIGURE 6.20
Open-loop root locus and Bode plots for practical compensated DC servo system.

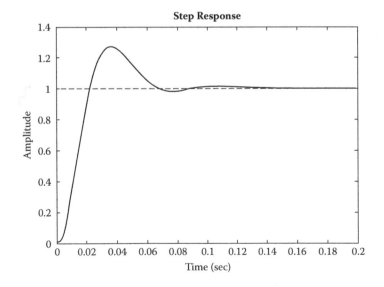

FIGURE 6.21
Closed-loop step response for practical compensation.

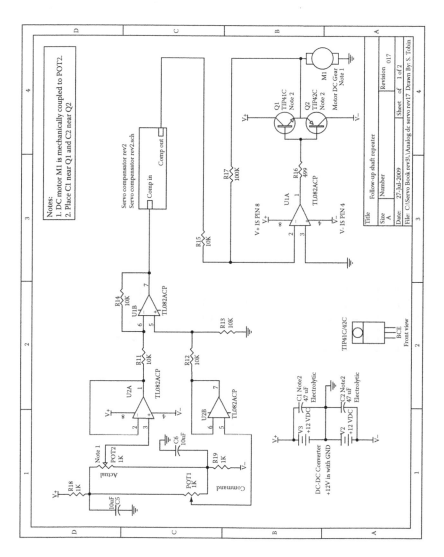

FIGURE 6.22

Final schematic for follow-up shaft repeater.

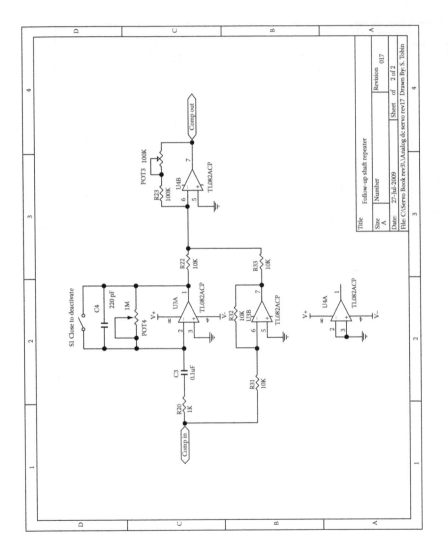

FIGURE 6.23
Shaft repeater compensation circuit schematic.

from –10 to –20. This is combined with U1A's overall gain of –10, giving a total span of 100 to 200. At the lowest gain, the servo will probably oscillate with S1 open. It is therefore advisable to start by keeping S1 closed and adjusting POT4 to its approximate midpoint. Then, S1 can be opened and transient response tuned by a gain adjustment (POT3) and a slight damping adjustment (POT4) if necessary.

References

Doebelin, E., *Dynamic Analysis and Feedback Control*, McGraw-Hill, 1962.

Franco, S., *Design with Operational Amplifiers and Analog Integrated Circuits*, McGraw-Hill, 1988.

Fredricksen, T., *Intuitive Operational Amplifiers*, McGraw-Hill, 1988.

Hayes, T., and P. Horowitz, *Learning the Art of Electronics*, Cambridge, 2010.

National Semiconductor, "An Applications Guide for Op Amps," AN-20, 1980.

Nise, N., *Control Systems Engineering*, Wiley, 2004.

Tietze, U., and Ch. Schenk, *Electronic Circuits: Design and Applications*, Springer-Verlag, 1991.

7

DC Servo Amplifiers and Shaft Encoders

7.1 Introduction

Two necessary elements of any DC servo system are an amplifier and a feed-back transducer. In this chapter we will take a look at the most common form of each element as used in fractional-horsepower DC servos today. These are (1) the pulse width modulated (PWM) switch-mode amplifier and (2) the optical incremental shaft encoder. We present these two elements together in the same chapter because the encoder provides digital (TTL-compatible) input to a microcontroller, and the PWM amplifier can take a fixed-frequency digital pulse train of varying duty cycle from a microcontroller and provide any average voltage between zero and 100 percent of rating to a brushed DC motor. These two innovations, used together in the 1980s by Hewlett-Packard in their high-performance plotter [Baron, 1981] and printer [Jackson, 1988] products, made microprocessor control of DC motors possible.

7.1.1 Scope of Discussion

As shown in Figure 7.1, both the optical shaft encoder and the PWM ampli-fier are "digital" devices. Specifically, the encoder takes analog shaft rota-tion as input and provides one or more digital pulse trains as an output. Two channels are often provided in quadrature for detecting direction of rotation. Each pulse represents a fixed, finite increment of angular rotation. The amplifier takes a digital input (fixed frequency, varying duty cycle pulse train) and provides a specific average analog voltage to the motor. These duty cycles are preset by a microcontroller in discrete steps, the number of steps depending on the resolution of the pulse generator located either inside or outside the microcontroller itself. Our plan will be to generate models for these components that fit into the continuous-time domain block diagrams we have been using throughout the book. This is an approximation that has been used with success by HP and others throughout the 1980s and 1990s. Of course, we are sidestepping the entire theory of discrete-time control and the Z-transform technique of handling sampled data mathematically. For our purposes in this book, the continuous approximation of these components is

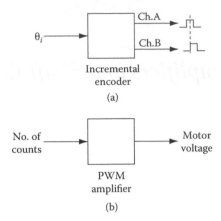

FIGURE 7.1
Simplified block diagrams of (a) incremental encoder and (b) PWM amplifier.

adequate. The formal theory of sampled data will not be used, except in very elementary ways when we look at implementing control laws (proportional, integral, derivative) within the firmware of a microcontroller.

We will look first at PWM amplifiers and then examine incremental shaft encoders.

7.2 DC Servo Amplifiers

To amplify an error signal and apply it to the terminals of a DC motor, we essentially have two choices. The first is a linear amplifier, and a simple example is given in the shaft repeater design presented earlier. The main advantage is simplicity and low parts count. The disadvantages are twofold: (1) the transistors operate in the linear region, dissipating excess power; (2) positive and negative power supplies are required to get bidirectional motion in a push-pull arrangement, increasing cost and decreasing efficiency. It is rare to see a DC motor driven this way unless the output power requirement is below 50 W. For the interested reader, a design example for a push-pull linear amplifier designed by R. Schmidt is given in [Kuo and Tal, 1978]. PWM amplifiers, on the other hand, are widely employed today, and we will examine these in detail now.

7.2.1 The Nature of PWM

[Taft and Slate, 1979] explains that PWM amplifiers are amplifiers that use transistors operating in the switching mode. By switching the transistors on into saturation and then off, power losses in the transistor junctions can be minimized. This means that potentially less expensive power transistors and smaller heat sinks can be used by the designer. Then, if the transistors are switched between two voltage levels, at a frequency well beyond the bandwidth of the driven system, the motor/load combination will act as a low-pass filter, responding only to the low-frequency components of the amplifier's output signal. Finally, when using switching amplifiers to drive inductive loads, it is necessary to protect the transistors by using so-called freewheeling diodes. In the bidirectional drive scheme the diodes are placed directly across the transistors. In many cases where MOSFET transistors are used, the diodes are intrinsic to the device structure.

7.3 PWM Switch-Mode Amplifiers

The most common configuration for driving brushed DC motors is an H-bridge. The main advantage of the H-bridge is that it allows bidirectional control using a single power supply, without having to resort to split supplies. Because our main goal in this book is to get an amplifier up and running quickly, we will be choosing an off-the-shelf solution based on the National Semiconductor LMD18200 integrated circuit. Prior to implementing this solution, we will investigate the main principles underlying the use of the H-bridge to drive DC motors.

7.3.1 H-Bridge Topology

The ideal circuit for an H-bridge is shown in Figure 7.2. It consists of four static, bidirectional switches, providing four-quadrant control to a load connected

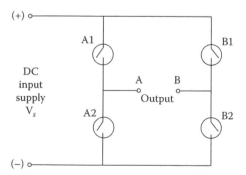

FIGURE 7.2
Ideal circuit for an H-bridge.

between points A and B. *Static* means the switches can remain in a given state (open or closed) for any time period we choose. *Four-quadrant* means the voltage across the load and the direction of current through the load can be of either polarity. For example, say we want a DC motor to rotate in one direction at approximately half of its rated no-load speed, with no external load applied to the output shaft, using an H-bridge as a driver. Suppose the direction of rotation we want in our setup is produced when current flows from terminal B to terminal A.

Assuming all switches are open at time $t = 0$, we proceed to close switches B1 and A2. This applies a step change in voltage across the motor windings of magnitude V_s. What does the motor current waveform look like? We know its magnitude will be small, because the current draw is proportional to the torque needed to accelerate its own rotor inertia, overcome any external load (which is zero in this case), and overcome any internal friction. Though small in magnitude, we know the current will increase, and the motor will begin to move. Now, to modulate the pulse width of the motor's applied voltage, we open switch A2 at time $t = t_1$, then close switch A1 at time $t = t_1 + \delta t$, where $\delta t \ll t_1$. This small delay δt prevents a power supply short circuit and so-called shoot-through current occurring with the short. As shown in Figure 7.3, the applied voltage drops to zero, and the current contained in the motor's inductance, which cannot change instantaneously, is given a path to flow in switches A1 and B1. During this time when the current circulates, it

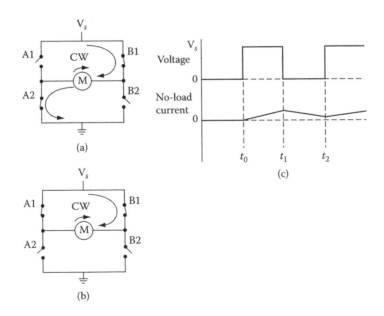

FIGURE 7.3
No-load H-bridge drive sequence for one rotation direction.

decreases due to the motor's back emf, which is present as long as the motor is moving, but opposes the flow of the original current during time $0 < t < t_1$. To repeat the PWM process, we open switch A1 at time $t = t_2$ and close switch A2 at time $t = t_2 + \delta t$. The voltage V_s is replaced across the winding and the current increases again. If we make the time periods 0 to t_1 and t_1 to t_2 equal, this produces a 50 percent duty cycle and the motor will rotate at approximately half of its rated no-load speed. Note that at time t_0 the current is zero; however, at time t_2 it returns to a small but nonzero value.

7.3.2 Waveform Analysis

To get bidirectional performance, let us give the bridge circuit of Figure 7.2 the following switching possibilities:

Case 1 (applied voltage positive)
B1 turned on continuously
A2 turned on during $0 \le t < t_1$
A1 turned on during $t_1 \le t < t_2$

Case 2 (applied voltage negative)
A1 turned on continuously
B2 turned on during $0 \le t < t_1$
B1 turned on during $t_1 \le t < t_2$

Persson [Kuo and Tal, 1978] refers to this switching scheme as "unipolar," whereas [Regan, 1990] calls it "sign/magnitude." It is one of two schemes Regan discusses, and its main advantage is that one of the upper switches is always on, giving a continuous current feedback signal from the LMD18200 IC. The resulting motor voltages for the two cases, shown in Figure 7.4, are given by

$$V_m = V_s; \text{ for } 0 \le t < t_1 \text{ (case 1)} \tag{7.1a}$$

$$V_m = -V_s; \text{ for } 0 \le t < t_1 \text{ (case 2)} \tag{7.1b}$$

$$V_m = 0; \text{ for } t_1 \le t < t_2 \text{ (both cases)} \tag{7.1c}$$

where

$$V_m \triangleq motor\ voltage,\ volt$$

$$V_s \triangleq H\ bridge\ supply\ voltage,\ volt$$

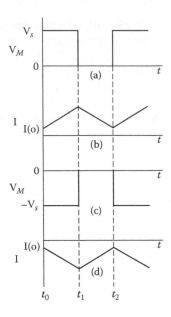

FIGURE 7.4
Voltage and current waveforms for sign/magnitude drive.

This square wave pattern for V_m can be expressed by a Fourier series:

$$V_m = a_0 + \sum_{n=1}^{\infty} a_n \cos(2\pi n f_s t + \varnothing_n) \qquad (7.2)$$

where

$$a_0 = \alpha V_s \qquad (7.3)$$

and

$$a_n = \left(\frac{2V_s}{n\pi}\right) \sin n\pi \, |\alpha| \qquad (7.4)$$

The switching frequency, f_s, is chosen to be much higher than the motor bandwidth and any resonance modes, so only the fundamental component of (7.2) needs to be considered. When averaged over many switching cycles,

$$V_m = \alpha V_s \qquad (7.5)$$

where

$$\alpha \triangleq load\ factor, \qquad 0 \le \alpha \le 1 \qquad (7.5a)$$

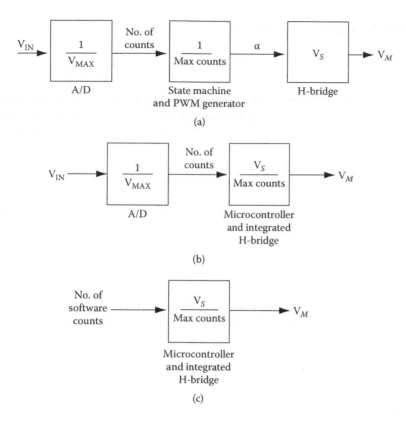

FIGURE 7.5
Switch-mode amplifier block diagram possibilities.

The duty cycle is simply

$$Duty\ cycle \triangleq \alpha \times 100\% \tag{7.6}$$

This switch-mode amplifier can then be represented by a choice of block diagrams as shown in Figure 7.5. If an analog-to-digital prescaling converter is used, we can define the input voltage to the A/D as V_{in} and the maximum value of V_{in} as V_{max}. A state machine is then used to convert the number of counts desired to a load factor. Very often, the functions of the A/D and state machine will be performed by software. In this case, the PWM signal is written directly to a microcontroller output port. A simple way to model the amplifier is [Jackson, 1988]

$$V_m = K_A(\#\ of\ counts) \tag{7.7}$$

The amplifier gain is expressed as

$$K_A = V_s / (max. \text{ # of counts}) \tag{7.8}$$

For example, in the world of 8-bit microcontrollers, the maximum number of counts might be 256, corresponding to $\alpha = 1$. Then each input count would correspond to a duty cycle of $1/256 \times 100$ percent $= 0.4$ percent. This is the resolution of the PWM amplifier. For 100 percent duty cycle, the "on" time t_1 is selected to be $t_1 = t_2$, giving a fully on amplifier (switching sequence case 1 above). For 0 percent duty cycle, $t_1 = 0$ and the amplifier is fully off. A third possibility, −100 percent duty cycle, gives a fully on amplifier with a negative voltage, and $t_1 = t_2$ again (switching sequence case 2 above). These times are related by

$$t_1 = |\alpha| t_2 \tag{7.9}$$

To analyze the current waveform, Persson examined the electrical equation of the motor as follows. For the electrical circuit, we have

$$i_a R_a + L_a \frac{di_a}{dt} + K_s \omega_o = V_m \tag{7.10}$$

where

$$i_a \triangleq armature\ current,\ amp$$

$$R_a \triangleq armature\ resistance,\ ohm$$

$$L_a \triangleq armature\ inductance,\ henry$$

$$K_e \omega_o \triangleq voltage\ drop\ due\ to\ back\ emf\ of\ motor,\ volt$$

$$K_e \triangleq motor\ back\ emf\ constant, \frac{volt}{\frac{rad}{sec}}$$

The solution of (7.10) is exponential; however, a straight-line approximation of the current vs. time can greatly simplify the analysis. First, make the assumption that the motor velocity is constant over a switching period. Then assume that the voltage drop contributed by the resistance is small compared with that contributed by the inductance. It is then reasonable to assume that the current flowing through the resistance portion of the circuit can be averaged over a switching period. We can now define the simplified variables

$$i_x = \left(\frac{1}{t_f}\right)\int_0^{t_f} i(t)\,dt \qquad (7.11)$$

and

$$V_x = R_a i_x + K_e \omega_0 \qquad (7.12)$$

Equation (7.10) then becomes

$$L\frac{di_a}{dt} = V_m - V_x \qquad (7.13)$$

Dividing through by L and integrating with respect to time, we get the solution for the current for case 1 (positive input voltage) being

$$I(t) = I(0) + \left(\frac{V_s - V_x}{L}\right)t; \text{ for } 0 \le t < t_1 \qquad (7.14a)$$

$$I(t) = I(t_1) - \left(\frac{V_x}{L}\right)(t - t_1); \text{ for } t_1 \le t < t_2 \qquad (7.14b)$$

For case 2 (negative input voltage) the solution is

$$I(t) = I(0) - \left(\frac{V_s + V_x}{L}\right)t; \text{ for } 0 \le t < t_1 \qquad (7.15a)$$

$$I(t) = I(t_1) - \left(\frac{V_x}{L}\right)(t - t_1); \text{ for } t_1 \le t < t_2 \qquad (7.15b)$$

The current $I(t)$ for both cases is shown graphically in Figure 7.4.

We might want to know how the current varies with the given variables in a sign/magnitude drive. During steady-state operation, the current is periodic, so we know that

$$I(t_2) = I(0) \qquad (7.16)$$

Using positive input voltage as an example, these quantities can be subtracted and set equal to zero, which results in

$$\left(\frac{V_s - V_x}{L}\right)t_1 - \left(\frac{V_x}{L}\right)(t_2 - t_1) = 0 \tag{7.17}$$

To obtain (7.17), Equation (7.14a) was evaluated at a time

$$t = t_1 - \delta t; \; \delta t \to 0 \tag{7.18}$$

Combining (7.17) with (7.9) we get after some algebra

$$\alpha = V_x / V_s \tag{7.19}$$

This says that the load factor is proportional to the simplified variable V_x over one switching period. The expression (7.19) becomes false as the motor voltage V_m is averaged over many cycles of switching and (7.5) takes its place. Staying with our linear approximation, the current variation over a switching period is

$$\Delta I = I(t_1) - I(0) \tag{7.20}$$

Again, we evaluate (7.14a) at a time indicated by (7.18). $I(0)$ is eliminated, and the result is

$$\Delta I = \left(\frac{V_s t_2}{L}\right)(|\alpha| - \alpha^2) \tag{7.21}$$

The maximum current variation can be found by differentiating (7.21) with respect to α and setting the result equal to zero:

$$1 - 2\alpha = 0 \tag{7.22}$$

The maximum variation occurs at $\alpha = 0.5$, giving

$$\Delta I_{max} = \frac{(V_s t_2)}{4L} \tag{7.23}$$

We find the "ripple current" ΔI to be proportional to the H-bridge supply voltage and the switching period, and inversely proportional to the motor's inductance.

7.3.3 Other Switching Schemes

Another common H-bridge switching method is known as "bipolar" or "locked antiphase." In this scheme, the diagonally opposed pairs of switches in Figure 7.2 (A1-B2 and A2-B1) are driven on and off together. At zero average output voltage, the average voltage at each output terminal is at $V_s/2$. For this condition, the switching command signal has a 50 percent duty cycle and the average motor current is zero. One hundred percent duty cycle corresponds to full power for one direction of motor rotation, whereas 0 percent corresponds to full power in the opposite direction. The main advantage of this scheme is that it seems tailor-made for fast direction reversals in high-performance servo applications. Power is automatically decreased as the need to change direction arises. We will not demand fast direction reversals for examples given in this book, so we choose sign/magnitude control. The reader interested in this type of performance may consult the references given at the end of the chapter.

7.4 Sign/Magnitude Control with the LMD18200

A block diagram for the LMD18200 IC is given in Figure 7.6. The on-chip input logic state machine provides the truth table shown in Table 7.1. For sign/magnitude control, states 1 and 3 are used for one direction of rotation, and states 2 and 3 are used for the other direction. The brake line is normally low and can be grounded if no emergency braking is required for the application. The idealized switching waveforms are shown in Figure 7.7.

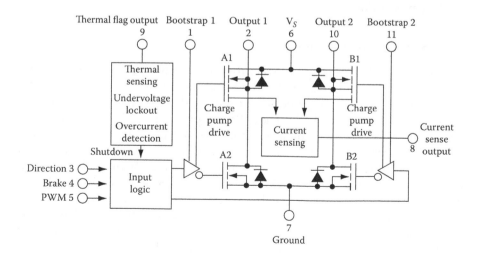

FIGURE 7.6
Block diagram for the LMD18200 IC (Regan, T., "A DMOS 3A, 55V, H-Bridge: The LMD 18200," National Semiconductor AN-694, 1990. Reproduced with permission of National Semiconductor Corporation).

TABLE 7.1

Control Logic Truth Table for the LMD18200 IC

State Number	PWM	Direction	Brake	Active Drivers	Comment
1	H	H	L	A1, B2	Sign/magnitude or locked antiphase drive
2	H	L	L	A2, B1	Sign/magnitude or locked antiphase drive
3	L	X	L	A1, B1	Sign/magnitude drive
4	H	H	H	A1, B1	Brake using upper switches
5	H	L	H	A2, B2	Brake using lower switches
6	L	X	H	NONE	No power applied

As seen in the figure, the current sense output is a continuous representation of the absolute value of the load current. This innovative current sensing circuitry does not rely on a small-valued resistor connected from the lower switches to ground. Instead, a few DMOS cells from each of the upper switches are separated out to provide a scaled replica of the total switch current. This scheme is shown in Figure 7.8. Only the forward current of each switch is sensed, since any reverse current would be blocked by the series diodes shown in the figure. The result is that as long as one of the upper switches is always active, the current will be continuously represented at pin 8. This signal is useful, for example, to actively limit the load current as a safety precaution.

7.4.1 Notes on Implementation

The author has driven a DC motor at different no-load speeds, using Program 15.4 written by [Sandhu, 2009] for the Micro Engineering Labs LAB-X1 PIC development board. The processor used is the Microchip Technologies PIC16F877A, clocked at 4 MHz. The software produces a 20 kHz PWM signal with duty cycle controlled by a potentiometer on the LAB-X1. As shown in Figure 7.9, three output ports on the microcontroller are connected to one of the two LMD18200 IC channels on board a Xavien 2-channel PWM amplifier module, part no. XDDCMD-1. The other LMD18200 IC on this module is unused. A Hansen DC motor rated at 12VDC with a 48:1 gear head was used, part no. 116-41232-48. This same general arrangement will be used later in the book, where an incremental encoder will be added to the motor's rear shaft extension to control its position. The oscilloscope photograph of Figure 7.10 shows the actual voltage waveform of the difference signal $(V_A - V_B)$ across the motor windings at approximately 50 percent duty cycle. Note that there is very little distortion due to the inductive "kick" normally found in bipolar

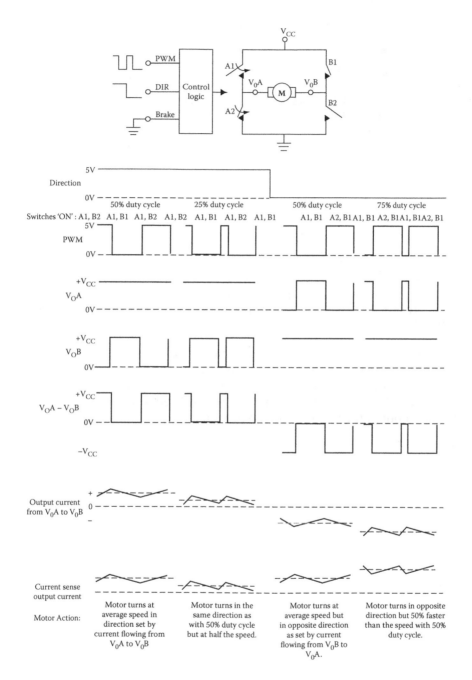

FIGURE 7.7
Sign/magnitude drive waveforms for the LMD18200 (Regan, T., "A DMOS 3A, 55V, H-Bridge: The LMD 18200," National Semiconductor AN-694, 1990. Reproduced with permission of National Semiconductor Corporation).

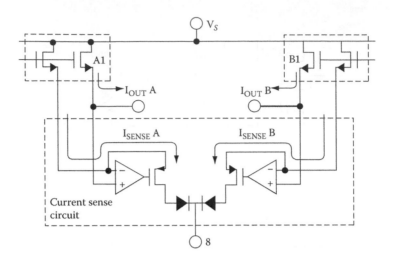

FIGURE 7.8
Current sensing circuitry of the LMD18200 (Regan, T., "A DMOS 3A, 55V, H-Bridge: The LMD 18200," National Semiconductor AN-694, 1990. Reproduced with permission of National Semiconductor Corporation).

transistor designs with freewheeling diodes [Barello, 2003]. Because the input signal to the motor is a square wave, the fundamental frequency as well as the higher harmonics can produce acoustical noise. The H-bridge fundamental frequency is normally chosen to keep any audible noise from the motor out of the range of human hearing.

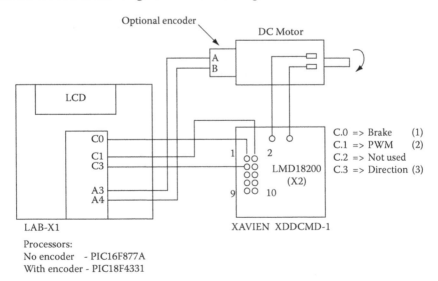

FIGURE 7.9
Digital control of a DC motor, block diagram.

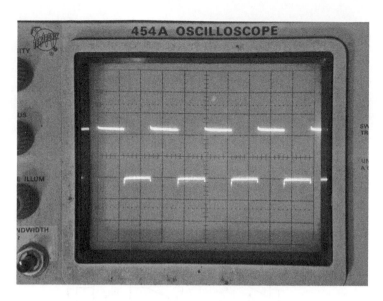

FIGURE 7.10

Voltage waveform across motor windings for sign/magnitude drive. (Oscilloscope scale is 5 V/div and 20 µs/div.)

7.5 Voltage Source versus Current Source

The PWM switch-mode amplifier just described functions as a voltage source. Its transfer function is expressed in units of V/V. It can also be arranged as a current source ("transconductance") amplifier. In this case, the transfer function is expressed in units of A/V. The push-pull linear amplifier can also be designed as either a voltage source or a current source, using the exact same principles. We will now take a look at each option in simplified form and decide when it might be beneficial to use one over the other in a particular application.

In Figure 7.11, (a) shows a voltage source and (b) shows a current source with a current limiting feature. Both circuits are similar in that they each have a pre-amplifier, intermediate amplifier, and power amplifier. The power amplifier of both circuits is connected within an op-amp feedback loop, which is a common practice for boosting the current capability of op-amp circuitry. When examining industrial schematics, the reader is advised to look for this pattern, where the power drive circuitry can contain many components, while the "signal level" circuitry can be integrated into just a few. Also, the power amplifier is shown as a push-pull type in both circuits. An H-bridge arrangement could be substituted into either one. Schmidt [Kuo and Tal, 1978] calls current limiting mandatory when driving DC motors

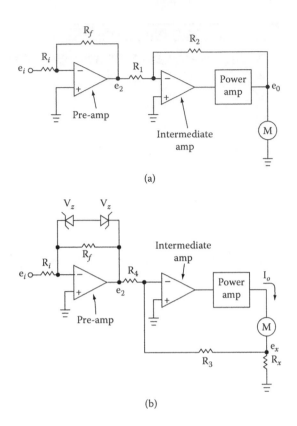

FIGURE 7.11
(a) Voltage source and (b) current source drive schematics.

with transistor amplifiers and cites the following characteristics of a current limiter as desirable:

1. Maintain relatively flat limiting characteristics even with large amounts of overdrive
2. Fast enough to prevent current overshoot when a motor is decelerating from high speed
3. Closed-loop in form to prevent large pre-amp error voltages when in current limit
4. Located as close as possible to the output transistors

We now proceed with an analysis of these two circuits and try to draw some useful conclusions. For the voltage source, the output voltage is simply

$$e_o = -e_2 \left(\frac{R_2}{R_1} \right) \qquad (7.24)$$

For the current source, we have two simultaneous equations, assuming a virtual ground at the negative input of the intermediate amplifier:

$$\frac{e_2}{R_4} + \frac{e_x}{R_3} = 0 \tag{7.25}$$

$$\frac{e_x}{R_x} + \frac{e_x}{R_3} = I_o \tag{7.26}$$

The sense resistor is usually selected to have a small voltage drop, so there is as much voltage available to drive the motor as possible. If R_x is much smaller than feedback resistor R3, the transconductance is approximately given by

$$\frac{I_o}{e_2} \cong -\frac{R_3}{R_4 R_x} \tag{7.27}$$

Because the voltage e_2 in Figure 7.11(b) is limited to

$$e_{2max} = \pm(V_Z + V_D) \tag{7.28}$$

the current is limited to

$$I_{lim} = \pm(V_Z + V_D)\frac{R_3}{R_4 R_x} \tag{7.29}$$

7.5.1 Voltage and Current Source Stability Assessment

To assess the relative stability of these two motor drive arrangements for use in position control, we can first derive the transfer functions of the intermediate and power amplifier combinations of Figure 7.11. Then these can be placed inside a position servo loop, and the root loci of each can be evaluated. Looking at Figure 7.12(a), for the current source we have the electrical equation

$$\frac{V_i}{R_4} + \frac{iR_x}{R_3} = 0 \tag{7.30}$$

The electromechanical relation in the Laplace domain is

$$K_T i(s) = (Js + B)\omega_o(s) \tag{7.31}$$

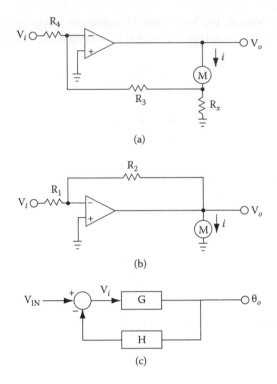

FIGURE 7.12
(a) Simplified current source, (b) simplified voltage source, and (c) position feedback control system.

Substituting (7.31) into (7.30) we get

$$\frac{\omega_o}{V_i}(s) = -\frac{R_3 K_T}{(Js+B)R_x R_4} \tag{7.32}$$

Because position is the integral of velocity, we insert another factor of s into the denominator and get

$$\frac{\theta_o}{V_i}(s) = -\frac{R_3 K_T}{s(Js+B)R_x R_4} \tag{7.33}$$

For the voltage source of Figure 7.12(b), the electrical equation is

$$\frac{V_i}{R_1} + \frac{V_o}{R_2} = 0 \tag{7.34}$$

The electromechanical relation in the Laplace domain is

$$V_o(s) = \left[K_e + \frac{(R+sL)(Js+B)}{K_T} \right] \omega_o(s) \tag{7.35}$$

Again substituting (7.35) into (7.34) we get

$$\frac{\omega_o}{V_i}(s) = -\frac{R_2 K_T}{R_1[K_e K_T + (Js+B)(R+sL)]} \tag{7.36}$$

In terms of position,

$$\frac{\theta_o}{V_i}(s) = -\frac{R_2 K_T}{sR_1[K_e K_T + (Js+B)(R+sL)]} \tag{7.37}$$

Consider a position control system as shown in Figure 7.12(c), where

$$\frac{\theta_o}{V_{in}} = \frac{G}{1+GH} \tag{7.38}$$

and

$$G = \frac{\theta_o}{V_i}(s) \tag{7.39}$$

for a current source (7.33) or a voltage source (7.37), and $H = 1$ for simplicity of analysis. We will also include the pre-amplifier of Figure 7.11, choosing a nominal gain of –1 to avoid a signal inversion (180° phase shift). The characteristic equation of (7.38) may be written as [Ogata, 2008]

$$1 + K\frac{num}{den} = 0 \tag{7.40}$$

where *num* and *den* are the numerator and denominator polynomials of the loop gain, *GH*, and *K* is a gain that is varied from zero to infinity. We can now use MATLAB's control system toolbox to plot the root loci of the current source and voltage source inside a position control loop. For example, we can compare a voltage source with a nominal gain of 1 V/V to a current source with nominal gain 1 A/V using the following code in MATLAB:

```
R  = 2.0;  % ohm
L  = 0.5;  % henry
Kt = .015;  % N-m/A
Ke = .015;  % V-sec/rad
J  = .02;  % kg-m^2
B  = 0.2;  % N-m-sec
```

```
R1=10e3; % ohm
R2=10e3; % ohm
R3=10e3; % ohm
R4=100e3; % ohm
Rx=0.1; % ohm
numi=R3*Kt;
deni=[J*Rx*R4 B*Rx*R4 0];
sys_dci=tf(numi,deni);
numv=R2*Kt;
denv=[J*L*R1 (J*R+B*L)*R1 (Kt*Ke+R*B)*R1 0];
sys_dcv=tf(numv,denv);
```

Figure 7.13 is the result of the "sisotool" command for the voltage source drive. It is stable up to a gain of about 50 dB, and higher if compensated appropriately. Figure 7.14 shows the same result for the current source drive, which is predicted to be stable for all gains. From a purely control system viewpoint, we could argue that the pole involving the inductance is eliminated in the case of the current source, accounting for its superior stability. In reality, we are looking at simplified analog implementations of these drive schemes. Because the current through an inductive load cannot change instantaneously, the best we can do is increase the initial slope (V/L) of the rising current waveform by increasing the drive voltage. The voltage on the load will rise linearly,

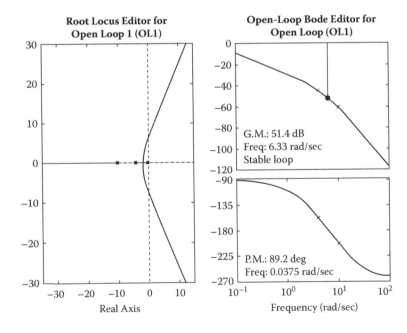

FIGURE 7.13
Root locus and Bode plots for voltage source drive.

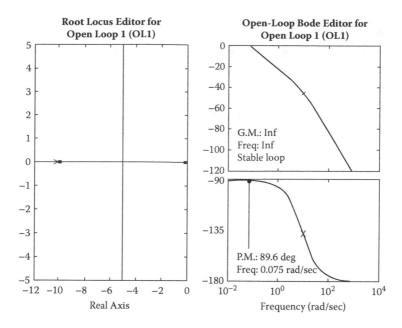

FIGURE 7.14
Root locus and Bode plots for current source drive.

up to the op-amp's saturation point, until the commanded current is attained. This amounts to decreasing one time lag in the overall system but not eliminating it. A more elaborate scheme for sourcing current, given by [Regan, 1990] and shown in Figure 7.15, is a "fixed off time" control loop, which is one example of chopper-type control. Chopper drives require high supply voltages, much higher than the motor's rating, to come close to eliminating the time lag due to the inductance. So we can also say that constructing a useful current source is more expensive and noisier than constructing a useful voltage source. The conventional wisdom is to use a compensated voltage source for controlling position or velocity and a current source for controlling torque or force. It is recommended that any compensated voltage source be equipped with current limiting, which is relatively easy to implement with the LMD18200. All that is needed is a resistor to ground and one of the A/D channels on a PIC microcontroller.

7.6 Shaft Encoders

As mentioned at the beginning of the chapter, the most common implementation of shaft encoders today in position servos is the optical incremental type. This is mainly due to high reliability (no moving parts contact each

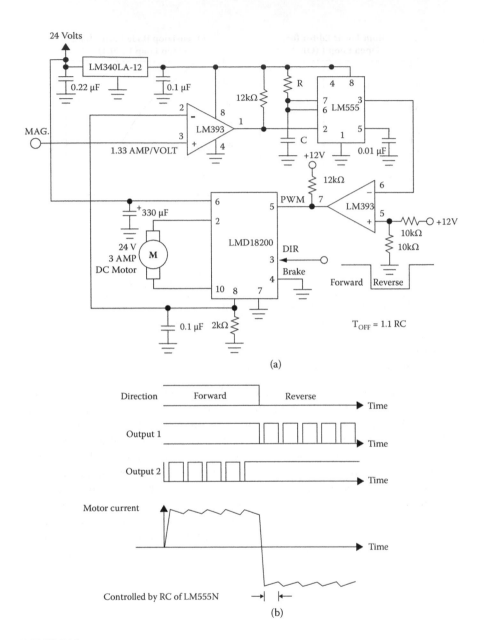

(a)

(b)

FIGURE 7.15
Fixed off time control, (a) schematic implementation and (b) switching waveforms (Regan, T., "A DMOS 3A, 55V, H-Bridge: The LMD 18200," National Semiconductor AN-694, 1990. Reproduced with permission of National Semiconductor Corporation).

other) and low cost. It is essentially a position transducer that reports the angle of shaft displacement in discrete steps. These encoders give only relative position information, so absolute references are needed in any real machine. These are known usually as "home" or "limit" sensors. One example is the carriage of an ink-jet printer, where "end of line" sensors must be supplied on both ends of the carriage travel mechanism to indicate to the motion controller when to reverse the direction of travel. Hewlett-Packard succeeded in combining these two functions using a linear encoder strip in 1988 [Ellement, 1988].

7.6.1 The Optical Rotary Incremental Encoder

The most common form of shaft encoder in use today is the optical rotary incremental type. To reduce inertia and eliminate resonant modes in the drive trains in which they are used, these units are usually supplied by manufacturers in so-called kit form. The kit encoder is typically assembled by the motor supplier onto the rear of the motor, which is designed with a rear shaft extension and tapped holes on the rear end cap. This arrangement is preferred because of its simplicity and low cost over encoders containing a shaft and bearings. A U.S. Digital 500-line kit encoder is shown attached to a Hansen DC gear motor in Figure 7.16. This encoder uses an HP/Avago HEDS-9000 series integrated LED emitter/photodiode detector module, which is designed to accept either rotary encoder wheels or linear encoder strips. A different HEDS-9000 module is required for each code wheel resolution, due to the technical and economic trade-offs made in its design [Epstein, 1988]. A U.S. Digital E2-500-187-I kit encoder is shown

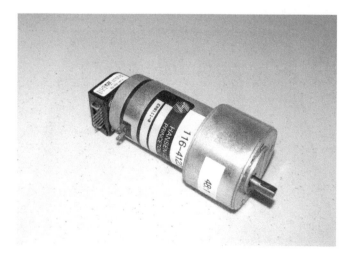

FIGURE 7.16
500-line kit encoder attached to a DC gear motor.

FIGURE 7.17
Main components of a 500-line kit encoder.

disassembled from one of the Hansen motors ordered by the author in
Figure 7.17.

After due consideration, it was decided to abandon the 500-line-resolution
encoder ordered from Hansen in favor of a lower-resolution 20-line device
supplied by EncoderGeek.com. This encoder is a very simple 2-channel
quadrature device, but it has all the essential features of mass-marketed units.
Separate code wheels, reader boards, and aluminum brackets were bought,
and the author retrofitted one of the Hansen motors with this encoder system. The essential parts are shown in Figure 7.18, and the retrofitted motor

FIGURE 7.18
Main components of a 20-line kit encoder.

FIGURE 7.19
DC gear motor retrofitted with a 20-line kit encoder.

is shown in Figure 7.19. A shim 0.062 in. thick was made from perforated FR-4 stock to be inserted between the bracket top and the reader board. The shim was required to accurately position the reader board relative to the code wheel openings.

7.6.2 Principle of Operation

Optical rotary incremental encoders sense shaft motion by (1) interrupting a pair of stationary light beams with the spokes of a single rotating code wheel and (2) using two stationary light detectors to receive a pattern of shadows. Each shadow pattern, when detected as an electrical signal, can be recorded against time as the motor spins to give a number of lines per second, which is constant for a given motor speed. A block diagram [Epstein, 1981] of the Hewlett-Packard HEDS-5000 type kit encoder, a precursor to the HEDS-9000 devices, is shown in Figure 7.20(a). This encoder broke new ground in terms of low cost, reliability, and high resolution and is covered under U.S. patent no. 4,266,125. The output signals from the encoder are similar digital waveforms. Each cycle of 360 electrical degrees corresponds to the angle between adjacent code wheel spokes. The idealized photocurrent waveforms of the photodiode pair for channel "A" are shown in Figure 7.20(b). These are processed by two transimpedance amplifiers and a comparator for each channel to give the two digital outputs shown in Figure 7.20(c).

On further inspection, Figure 7.20(a) reveals a few subtle yet important details necessary for an optical encoder with push-pull-type outputs. First, the photodiode detectors and lenses covering them are spaced one code wheel pitch apart. Second, the openings in the phase plate for each channel

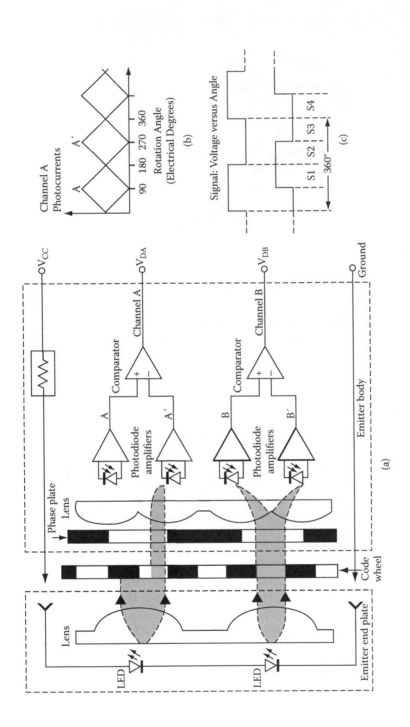

FIGURE 7.20

(a) Block diagram of HP HEDS-5000 type kit encoder, (b) idealized photocurrent waveforms for channels "A," and (c) digital output waveforms for channels "A" and "B." (Epstein, H. et.al., "An Incremental Optical Shaft Encoder Kit with Integrated Optoelectronics," Hewlett-Packard Journal, Oct. 1981. Reproduced with permission of Hewlett-Packard Co. as given in http://www.hpl.hp.com/hpjournal/permission.htm, © 1981 Hewlett-Packard Co.)

are twice the width of the code wheel openings. Taking the upper opening as an example, the upper half is 180° out of phase with the code wheel (light is blocked), while the lower half is in phase with the code wheel (light is transmitted). This arrangement is duplicated at the lower opening, producing a 180° phase shift between detected signals A and A' and between detected signals B and B'. Third, the phase plate openings covering channel A and channel B are offset by one code wheel pitch plus an extra ¼-pitch. This ¼-pitch displacement produces a 90° phase shift between processed signals V_{DA} and V_{DB}, the D subscript indicating *digital*. Electrically this relationship is called quadrature, and it allows the user to decode the direction of code wheel rotation as shown in Figure 7.20(c). For one direction (left to right as pictured), the logic states of the two channels occur in the sequence S1, S2, S3, S4. For the opposite direction, they occur in the sequence S1, S4, S3, S2. One basic method of hardware direction decoding using sequential logic gates (*D*-flip-flops) is shown in Figure 7.21. Other more elaborate schemes decode direction and increase resolution by a factor of 2 or 4 and can be purchased in hardware ICs or programmed in firmware using small microcontrollers [Harrison, 2005].

The 20-line EncoderGeek.com device is a simplified version of the aforementioned schemes. The push-pull output feature is not required, which was instituted by HP to compensate for changes in LED brightness over long time periods. The photodiodes and transimpedance amplifiers are replaced by phototransistors, one for each channel, so the device is less expensive. LED/lens and phase plate/lens are replaced by the components of two

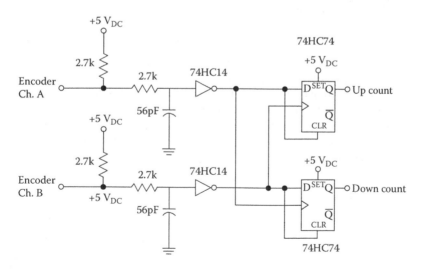

FIGURE 7.21
Hardware direction decoding using *D*-flip-flops.

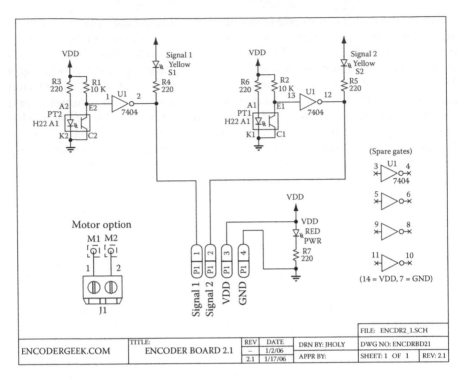

FIGURE 7.22
Schematic of 20-line encoder (Epstein, H. et.al., "An Incremental Optical Shaft Encoder Kit with Integrated Optoelectronics," Hewlett-Packard Journal, Oct. 1981. Reproduced with permission of EncoderGeek.com).

Motorola H22A1 photo interrupters. The only stipulation is that the H22A1s must be displaced by a full code wheel pitch plus an extra ¼-pitch offset. The encoder schematic, given in Figure 7.22, shows an active pull-down when the phototransistors are exposed to light, followed by an inverter gate. This was updated by the vendor to a 74F14 Schmitt trigger inverter, whose output acts like a comparator with built-in hysteresis. Extra LEDs are added to the top side of the board to indicate the status of each channel at minimal cost. For low resolutions, use of the photo interrupters is an elegant solution.

7.6.3 Signal Transfer through Cables

The author chose to connect the Encoder Geek device to the LAB-X1 board using a common flat 4-conductor unshielded telephone cable. The header on the encoder reader board has the two channel outputs on adjacent pins. If the channel signals are adjacent to each other throughout a 1 m length of the telephone cable, there will be enough capacitive coupling between them to compromise signal integrity. A simple solution is shown schematically

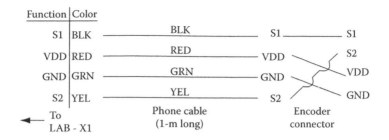

FIGURE 7.23
Cabling schematic to reduce capacitive coupling between channels.

in Figure 7.23, where the outer two conductors carry the output signals, and the inner two conductors carry VDD and GND. This reduced the coupling sufficiently for all encoded motor work in this book to be successfully implemented.

References

Barello, L., "H-Bridges and PMDC Motor Control Demystified," http://www.barello.net/papers/index.htm, 2003.

Baron, W. et al., "Development of a High-Performance, Low-Mass, Low-Inertia Plotting Technology," *Hewlett-Packard Journal*, Oct. 1981.

Ellement, D., and M. Majette, "Low-Cost Servo Design," *Hewlett-Packard Journal*, Aug. 1988.

Epstein, H. et al., "An Incremental Optical Shaft Encoder Kit with Integrated Optoelectronics," *Hewlett-Packard Journal*, Oct. 1981.

Epstein, H. et al., "Economical, High-Performance Optical Encoders," *Hewlett-Packard Journal*, Oct. 1988.

Harrison, P., "Quadrature Decoding with a Tiny AVR," http://www.helicron.net/avr/quadrature, 2005.

Jackson, L. et al., "Deskjet Printer Chassis and Mechanism Design," *Hewlett-Packard Journal*, Oct. 1988.

Kuo, B. and J. Tal, *Incremental Motion Control: DC Motors and Control Systems*, SRL, 1978.

Ogata, K., *Matlab for Control Engineers*, Pearson Prentice-Hall, 2008.

Regan, T., "A DMOS 3A, 55V, H-Bridge: The LMD 18200," National Semiconductor AN-694, 1990.

Sandhu, H., *Running Small Motors with PIC Microcontrollers*, McGraw-Hill, 2009.

Taft, C., and E. Slate, "Pulsewidth Modulated DC Control," *IEEE Trans. Ind. Elec. and Control Inst.*, vol. IECI-26, Nov. 1979.

8

Control of a Position Servo Using
a PIC Microcontroller

8.1 Introduction

Suppose we are given data from a computer that represents the design of a new home as developed by an architect. The data consists of black lines on a white background. Some of the lines are straight, some are curved, and some are at different angles relative to the border of the drawing. Many of the lines are used to form readable characters. Our assumption is that the design of the home is "computer aided" and that the architect is no longer interested in producing drawings using a pencil and paper (or vellum) by hand. If a machine could produce these drawings, the architect could make a new copy each time a change is made. This would also allow for the automated production of many copies of the same drawing for distribution to builders, vendors, and clients.

Further suppose that we are interested in the design of a machine that would take computerized data input by the architect and either "draw" or "print" the drawing represented by this data in a relatively short time period. We can imagine that to produce a drawing automatically, some form of motion of pen or pencil relative to paper is required. Furthermore, to faithfully reproduce a drawing several times, we can say that the motion needs to be controlled in some way. As we have already demonstrated, one way motion can be controlled is with a DC servo. A significant challenge we face is that the servo input (commanded position) is now in some kind of digital format instead of a continuous analog signal.

8.1.1 On-the-Fly versus Preprogrammed Moves

There are two major machine types that have been responsible for producing drawings from computerized data over the past 30 years. The first is the two-axis ("X-Y") pen plotter, and the second is the two-axis ink-jet printer/plotter. Both of these technologies were pioneered by Hewlett-Packard, the pen plotter in the 1980s and the ink-jet printer in the 1990s. The pen plotter

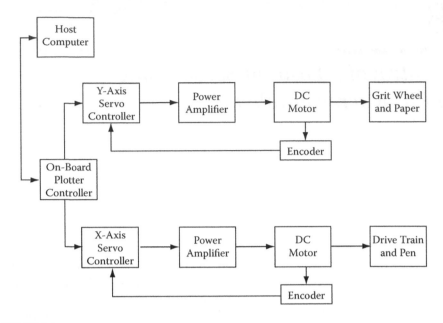

FIGURE 8.1
HP pen plotter schematic diagram. (Baron, W. et.al., "Development of a High-Performance, Low-Mass, Low-Inertia Plotting Technology," Hewlett-Packard Journal, Oct. 1981. Reproduced by permission of Hewlett-Packard Co. per http://www.hpl.hp.com/hpjournal/permission. htm, © 1981 Hewlett-Packard Co.)

is no longer popular, as it has been replaced by the ink-jet printer; however, we can use it to illustrate the difference between an on-the-fly move and a preprogrammed move [Baron, 1981].

For example, the HP X-Y pen plotter used one axis to move the pen and the other axis to move the paper, as shown schematically in Figure 8.1. However, there were two major disadvantages to the pen plotter: (1) speed was limited by the flow of ink to the tip of the pen, and (2) both axes had to be precisely coordinated with each other to form the correct image on the paper. This axis coordination implies that to draw a straight line correctly at some angle relative to the paper edge, the speeds of the axes always needed to be related by the same constant value throughout the move. Not only was position control required, but also velocity control. The control loop as modeled by HP is shown in Figure 8.2, and the line drawing process is illustrated in Figure 8.3. Trapezoidal velocity profiles were used, because slowing the motion down at the end of the move makes the resulting position response more stable, in exchange for a small time penalty. An infinite combination of moves could be sent to the plotter (one at a time), so that onboard calculation of position as well as continuous estimation of velocity was required using a single transducer on each axis, an optical incremental encoder. From one encoder pulse to the next, velocity was estimated by dividing the position information by the time between pulses. This in turn required the servo sampling rate to be

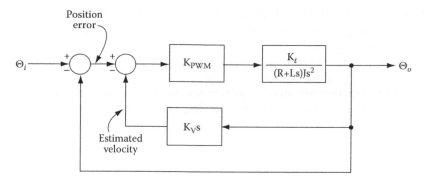

FIGURE 8.2

Pen plotter control loop as modeled by HP. (Baron, W. et.al., "Development of a High-Performance, Low-Mass, Low-Inertia Plotting Technology," Hewlett-Packard Journal, Oct. 1981. Reproduced by permission of Hewlett-Packard Co. per http://www.hpl.hp.com/hpjournal/permission.htm, © 1981 Hewlett-Packard Co.)

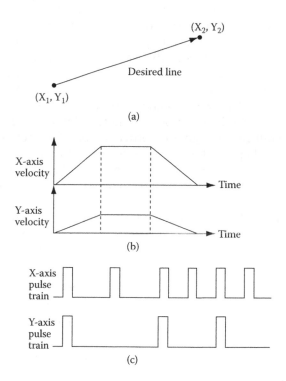

FIGURE 8.3

Pen plotter line drawing process. (Baron, W. et.al., "Development of a High-Performance, Low-Mass, Low-Inertia Plotting Technology," Hewlett-Packard Journal, Oct. 1981. Reproduced by permission of Hewlett-Packard Co. per http://www.hpl.hp.com/hpjournal/permission.htm, © 1981 Hewlett-Packard Co.)

at least twice as high [Leigh, 1992] as the maximum encoder pulse rate. The servo sampling rate is determined by how many "ticks" from a clock with a known frequency need to be counted between encoder pulses to accurately estimate velocity, calculate the next incremental position command, and provide a usable damping term for the position control loop.

Contrast this complex set of requirements for the X-Y plotter with an automatic paper towel dispenser. The towel dispenser is a single-axis device that makes the same move each time it is commanded by the tripping of a motion sensor. To minimize the chance of a paper jam, we might want the motion to be based on a position profile rather than a timer alone. For example, a finite set of moves can be preprogrammed into the paper towel dispenser, one for each length of paper to be dispensed. In contrast, the X-Y plotter was required to calculate the trajectory of each move sent to it on the fly.

8.1.2 Scope of Discussion

If a move is to be preprogrammed, its trajectory (position profile versus time) must be known well in advance of its execution. Conversely, the trajectory of an on-the-fly move may not be known until perhaps milliseconds before it is to be made. This makes the equipment necessary for an on-the-fly system more complex. Example block diagrams representing these systems are shown in Figure 8.4. Device (a) requires a dedicated host computer to generate software move commands in near real time, while device (b) is controlled by embedded firmware. Table 8.1 is intended to give the reader a qualitative feel for the features of each device, based on metrics important in discrete-time control. Our overall goal is to give a useful example of discrete-time DC servo control in a single chapter. This can be done only if the system is not too complex. We can limit complexity but at the same time illustrate all the important trade-offs involved. Our strategy will be to take the following actions:

1. Concentrate on the control of a single motor, realizing that a multi-axis move is nothing more than driving two or more motors together, each with its own trajectory, on a common time schedule.

2. Gear the motor down, such that a relatively coarse encoder can be used to command a smaller output shaft motion per line, as well as provide a more useful torque/speed combination.

3. Preprogram a move of a particular trajectory that might be used in an ink-jet printer paper feed system or automatic paper towel dispenser.

4. Dismiss the use of derivative control, because of difficulties encountered with calculating velocity during each sample period and the limitation of unsigned math.

5. Examine the move at end of travel to look for any instability, since derivative control will not be used.

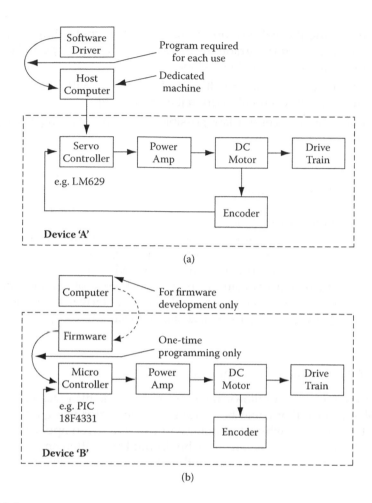

FIGURE 8.4
Schematics of equipment necessary for generating (a) "on-the-fly" and (b) preprogrammed move trajectories.

TABLE 8.1

Features of On-the-Fly versus Preprogrammed Digital Servos

Metric	On-the-Fly (Device A)	Preprogrammed (Device B)
Mathematics	Signed, floating point	Unsigned, fixed point
Precision	Higher	Lower
Processor speed	Faster	Slower
Software platform	Lower level with assembler	Higher level with compiler
Velocity feedback	Computed for each sample	Assumed from position profile
Control law	P-D control with nonlinear friction compensation	P-control with nonlinear friction compensation
Position counting	Done in software	Done in hardware

6. Use a triangular velocity profile, so that when controlling position the corrective effort is gradually increased on the up-slope and gradually decreased on the down-slope.

7. Allow the applied motor voltage to be nonzero with zero error to compensate for Coulomb friction [Sandhu, 2009].

8. Use the simplest control law possible, while maintaining adequate stability.

9. Reserve about 50 percent of the available power for driving an unknown load, and use the rest for overcoming friction and accelerating the rotor and gears.

As stated in the previous chapter, we will keep in mind the models for all components used that fit into the continuous-time domain block diagrams we have been using thus far. This kind of thinking has been used successfully by HP and others throughout the 1980s and 1990s. For those readers interested in further investigation, the formal theory of sampled data for implementing control laws (proportional, integral, derivative) is considered by [Kuo, 1995], [Leigh, 1992], [Atmel, 2006], and [Valenti, 2004].

8.1.3 DC Servos versus Step Motors

In the presentation that follows, the move to be made by a DC servo closely resembles a move that could be made by a step motor using a micro-step driver. The choice of which technology to use in a given application is not clear-cut by any means, and an entire book could be written on the subject of step motor systems. In general, the author has found the following guidance to be useful:

1. If external disturbances cannot be predicted, the step motor may stall and fail to complete a given move.

2. The audible noise produced by a DC servo is usually less than that produced by the step motor.

3. The DC servo is more efficient as a prime mover than the step motor.

4. The step motor is easier to implement than the DC servo because no encoder is required.

5. Usually the DC servo is used for higher-volume, cost-sensitive applications.

6. The step motor is often used for one-off laboratory applications where disturbances are predictable.

For further information on step motor systems, the interested reader may consult [Kuo, 1978], [Stiffler, 1992], [Hunt, 1993], and [Sandhu, 2009].

8.2 Initial Motor Selection

A DC gear motor was selected that is capable of doing a useful amount of work in a typical office machine or small industrial application. A motor capable of handling 25 W continuous input power seemed like a good compromise for use as an example in this book. A mid-range gearbox was selected with a 48:1 ratio, giving approximately 125 oz-in. of torque with 75 rpm speed at the output shaft, for 12 V, 0.5 A applied to the input terminals. To get an encoder-compatible rear shaft extension, the motor was ordered with an HP/Avago 500-line-per-revolution kit-type encoder preinstalled. Five hundred lines seemed to be in excess of what was needed, but it is by far the most popular count, making it the easiest to get without special ordering. A 12VDC gear motor with encoder was ordered from Hansen Corporation under part number 116-41232-48, with the encoder HP/Avago HEDS5540#A05 named as a separate line item on the purchase order. Hansen delivered the motor with U.S. Digital's version of the HP encoder, part number E2-500-187-I. A photograph of the motor-encoder combination as delivered by Hansen is shown in Figure 8.5. The rear shaft diameter for the encoder wheel is 0.187 inch, and the output shaft diameter is 0.250 inch.

The U.S. Digital 500-line encoder was removed and replaced with a 20-line-per-revolution kit encoder purchased from EncoderGeek.com. With the gear-reduced motor, the equivalent resolution at the output shaft was 960 lines per revolution, which seemed more than adequate for the example to be

FIGURE 8.5
Motor/encoder combination as delivered by Hansen Corp.

FIGURE 8.6
Motor/encoder combination with encoder replaced.

presented here. A photograph of this encoder mounted on the Hansen motor is shown in Figure 8.6.

8.3 Setting the Move Requirements

Consider a move of one revolution of the output shaft of our gear motor. Because the gear ratio is 48:1 and the encoder resolution is 20 lines per revolution of the input shaft, the move length is 960 lines. The motor-encoder combination was driven open-loop at 100 percent duty cycle using the ME Labs LAB-X1, Xavien amplifier, and Sandhu's program 15.4. The no-load speed was measured by looking at the encoder pulse train versus time on a Tektronix 454A oscilloscope and found to be about 4800 rpm. To be conservative, suppose we take the top speed to be 4000 rpm. This equates to 1.38 r/sec of the output shaft. Choosing a total move time of 2 sec with no load will use only about 50 percent of the power available, leaving the remainder to accelerate an unknown load.

8.3.1 The PIC18F4331 Quadrature Encoder Interface

To close the position loop, it is necessary to know how the encoder pulses are counted. As Sandhu points out, the selection of the 18F4331 microcontroller gives us the flexibility to know at all times what the actual position count is, just by looking at two dedicated 8-bit registers on this chip at any time. These registers are labeled POSCNTL (lowest significant byte) and POSCNTH (highest significant byte). These combine to form a 16-bit register, capable of a total count of 65,535. The translation from "lines" to "counts" is done by configuring bits 4 to 2 of the QEICON register. For this example, the QEI is enabled in "2X" update mode, with a reset occurring when POSCNT = MAXCNT (010 bit sequence). MAXCNT will not be used, because we will be resetting POSCNT explicitly in software. In 2X update mode, the position counter is clocked on both the rising and falling edges of the QEA encoder input. The QEB input is used to detect direction only. In the final analysis, one encoder line is equal to two encoder counts, making one revolution of the output shaft equal to 1920 counts. In programming the position profile, the main limitation is the necessity to always command an integer multiple number of counts at each sampling interval. This can serve as a check on the selected resolution for the encoder.

8.3.2 Velocity and Position Profiling

For a relatively short move, the simplest velocity profile is triangular. It gives constant acceleration on the up-slope and constant deceleration of equal magnitude on the down-slope. This allows for a predictable torque requirement throughout the move when driving inertial loads. We decided to make a move of 1920 counts in a time of 2 sec. A triangular velocity profile is given in Figure 8.7 for this move. The average velocity is simply the total distance traveled divided by the total time taken for the move:

$$V_{av} = 960 \ counts/sec \tag{8.1}$$

The distance traveled, s, is the integral of (or area under) the velocity profile curve, which in this case is a triangle:

$$s = \left(\frac{1}{2}\right)bh = \left(\frac{1}{2}\right)tV_{pk} \tag{8.2}$$

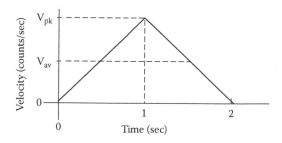

FIGURE 8.7
Triangular velocity profile for 2 sec move.

Solving for the peak velocity we get

$$V_{pk} = \frac{2s}{t} = 1920 \; counts/sec \qquad (8.3)$$

The specifications for the move to be made are

$$s \triangleq distance \; moved, \; counts = 1920$$

$$V_{av} \triangleq average \; velocity, \frac{counts}{sec} = 960$$

$$V_{pk} \triangleq peak \; velocity, \frac{counts}{sec} = 1920$$

$$t \triangleq move \; time, \; sec = 2$$

The desired position profile is made up of two quadratic segments meeting at the middle of the move and is illustrated in Figure 8.8.

8.3.3 Setting the Servo Sampling Rate

The basic idea behind a digitally controlled servo is to divide the move to be made into small, equally sized, discrete packets of time. This allows the programmed shaping of velocity profiles. A stable and accurate time base is required, and the simplest approach is to use the instruction cycle oscillator on board the LAB-X1. We can derive the sampling frequency by using the timer and interrupt features of the PIC18F4331. Sandhu used an overflow of Timer0 to generate an interrupt at equal time intervals. We will do the same thing here, selecting the lowest sampling rate that can do the job at hand. By selecting a low rate, we can ensure that the microcontroller will always get

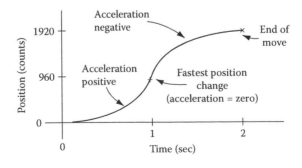

FIGURE 8.8
Position profile for 2 sec move shown in Figure 8.7.

through its entire program at least once before the next interrupt. This type of control requires us to ask for an integer number of counts at each interrupt. At the same time, we need the rate to be high enough for the control scheme to react to any system dynamics it encounters. For example, suppose the two control registers involved are set as follows:

$$INTCON = \% \ 10101100 \tag{8.4}$$

$$T0CON = \% \ 10000000 \tag{8.5}$$

From (8.4) and the PIC18F4331 data sheet [Microchip, 2007], the interrupt control register is set to use a Timer0 overflow to generate an interrupt. From (8.5), the Timer0 control register sets the timer to 16-bit, and its prescale value is set to 1:2 using the internal instruction cycle clock period (T_{cy}) as a reference. With these settings the servo sampling rate calculation is as follows:

Sampling rate calculation A

Timer0 is 16-bit = 65,536 counts

System clock frequency = 20,000,000 Hz

Instruction cycle clock frequency = $F_{osc}/4 = 1/T_{cy} = 5,000,000$ Hz

Prescale value = 1:2

Interrupt is enabled after $2 \times 65,536 = 131,072$ counts

Interrupt frequency = 5,000,000/131,072 = 38 Hz (this is the sampling rate)

Interrupt period = 1/38 = 0.0262 sec

Our strategy will be to update the commanded position after every interrupt period. For a total move time of 2 sec, we need to have the following number of updates:

$$Number \ of \ updates = \frac{2}{.0262} = 76.3 \ interrupts \tag{8.6}$$

Because the move consists of two equal segments, an even number of interrupts is required. We round (8.6) to 76 interrupts and adjust the total move time slightly to

$$Move \ time = 76(.0262) = 1.9912 \ sec \tag{8.7}$$

This results in each half of the move consisting of 38 interrupts and taking 0.9956 sec to complete.

We can estimate the upper bound of frequency content (bandwidth) of this servo as a function of the electrical time constant. Our assumption earlier in the book was that for a motor of this size,

$$\tau_e = \frac{L}{R} = 0.01 \; sec \tag{8.8}$$

The bandwidth is then

$$BW = \frac{1}{2\pi\tau_e} = 16 \; Hz \tag{8.9}$$

Shannon's theorem [Leigh, 1992] requires a sampling rate at least twice as fast as the quickest dynamic expected in the system. Our sampling rate of 38 Hz is just adequate. A rate closer to ten times the servo bandwidth (160 Hz) would be more desirable. Performing a second sampling rate calculation, we get

Sampling rate calculation B

Timer0 is 8-bit = 256 counts

System clock frequency = 20,000,000 Hz

Instruction cycle clock frequency = $F_{osc}/4 = 1/T_{cy} = 5,000,000$ Hz

Prescale value = 1:128

Interrupt is enabled after $128 \times 256 = 32,768$ counts

Interrupt frequency = 5,000,000/32,768 = 152.6 Hz (this is the sampling rate)

Interrupt period = 1/152.6 = 0.0065 sec

Updating the commanded position after every interrupt period, we have the following number of updates:

$$Number \; of \; updates = \frac{2}{.0065} = 307.7 \; interrupts \tag{8.10}$$

Because the move consists of two equal segments, an even number of interrupts is required. We round (8.10) to 308 interrupts and adjust the total move time slightly to

$$Move \; time = 308(.0065) = 2.002 \; sec \tag{8.11}$$

This results in each half of the move consisting of 154 interrupts and taking 1.001 sec to complete.

8.3.4 Calculating the Position Profile

Using the selected 20-line-per-revolution encoder, our goal is to move 1920 counts in a 2 sec time period. To preprogram this move with a triangular velocity profile, the commanded position at each interrupt must be known. This can be done using an Excel spreadsheet, with the help of the basic dynamic equations [Halliday and Resnick, 1977]

$$v_x = v_{x0} + a_x t \tag{8.12}$$

$$s = x_0 + v_{x0} t + \left(\frac{1}{2}\right) a_x t^2 \tag{8.13}$$

When the distance s is expressed in counts, the x coordinate can be thought of in terms of rectilinear motion, even though we are actually producing rotation of a shaft. The resulting position profile is given in Table 8.2.

8.3.5 Other Encoder Resolutions

The encoder resolution required to sample the servo at other rates can be calculated using the aforementioned equations. For example, for our sampling rate calculation B, the rate is 152.6 Hz. The first position command must be greater than or equal to 1 count. Knowing the time step size to be 0.0065 sec, we can use (8.13) to solve for the required acceleration, where

$$x_0 = 0 \tag{8.14a}$$

$$v_{x0} = 0 \tag{8.14b}$$

$$a_x = \frac{2s}{t^2} \tag{8.14c}$$

For $s = 1$ count, the acceleration is $2 \times 1/(.0065)^2 = 47,337$ counts/sec^2. Dividing the same 2 sec move into two equal time intervals, the number of counts required in 1 sec is

$$s = \left(\frac{1}{2}\right) a_x t^2 \tag{8.15}$$

Here, $t = 1$ sec and the distance required is 23,669 counts. To make a full revolution of the output shaft in 2 sec, 48 revolutions of the input shaft must take place. In 1 sec we need 24 revolutions, so that the encoder resolution needed is about 1000 counts. With the same QEI settings, a 500-line encoder seems almost ideal. As stated previously, this is a resolution kept in stock by many distributors of the HP/Avago shaft encoder product line.

TABLE 8.2

Position Profile for 76-Step Move

Step Number	Time (sec)	Velocity (counts/sec)	Position (counts)	Position (rounded)	Step Number	Time (sec)	Velocity (counts/sec)	Position (counts)	Position (rounded)
0	0	0	0	0					
1	0.0262	50.304	0.658982	1	39	0.0262	1869.696	1009.645	1010
2	0.0524	100.608	2.63593	3	40	0.0524	1819.392	1057.972	1058
3	0.0786	150.912	5.930842	6	41	0.0786	1769.088	1104.981	1105
4	0.1048	201.216	10.54372	11	42	0.1048	1718.784	1150.672	1151
5	0.131	251.52	16.47456	16	43	0.131	1668.48	1195.045	1195
6	0.1572	301.824	23.72237	24	44	0.1572	1618.176	1238.101	1238
7	0.1834	352.128	32.29014	32	45	0.1834	1567.872	1279.838	1280
8	0.2096	402.432	42.17487	42	46	0.2096	1517.568	1320.257	1320
9	0.2358	452.736	53.37757	53	47	0.2358	1467.264	1359.358	1359
10	0.262	503.04	65.89824	66	48	0.262	1416.96	1397.142	1397
11	0.2882	553.344	79.73687	80	49	0.2882	1366.656	1433.607	1434
12	0.3144	603.648	94.89347	95	50	0.3144	1316.352	1468.755	1469
13	0.3406	653.952	111.368	111	51	0.3406	1266.048	1502.584	1503
14	0.3668	704.256	129.1606	129	52	0.3668	1215.744	1535.095	1535
15	0.393	754.56	148.271	148	53	0.393	1165.44	1566.289	1566
16	0.4192	804.864	168.6995	169	54	0.4192	1115.136	1596.165	1596
17	0.4454	855.168	190.4459	190	55	0.4454	1064.832	1624.722	1625
18	0.4716	905.472	213.5103	214	56	0.4716	1014.528	1651.962	1652
19	0.4978	955.776	237.8926	238	57	0.4978	964.224	1677.883	1678
20	0.524	1006.08	263.593	264	58	0.524	913.92	1702.487	1702
21	0.5502	1056.384	290.6112	291	59	0.5502	863.616	1725.773	1726
22	0.5764	1106.688	318.9475	319	60	0.5764	813.312	1747.741	1748
23	0.6026	1156.992	348.6017	349	61	0.6026	763.008	1768.39	1768

24	0.6288	1207.296	379.5739	380
25	0.655	1257.6	411.864	412
26	0.6812	1307.904	445.4721	445
27	0.7074	1358.208	480.3982	480
28	0.7336	1408.512	516.6422	517
29	0.7598	1458.816	554.2042	554
30	0.786	1509.12	593.0842	593
31	0.8122	1559.424	633.2821	633
32	0.8384	1609.728	674.798	675
33	0.8646	1660.032	717.6318	718
34	0.8908	1710.336	761.7837	762
35	0.917	1760.64	807.2534	807
36	0.9432	1810.944	854.0412	854
37	0.9694	1861.248	902.1469	902
38	0.9956	1911.552	951.5706	952

62	0.6288	712.704	1787.722	1788
63	0.655	662.4	1805.736	1806
64	0.6812	612.096	1822.432	1822
65	0.7074	561.792	1837.81	1838
66	0.7336	511.488	1851.87	1852
67	0.7598	461.184	1864.612	1865
68	0.786	410.88	1876.036	1876
69	0.8122	360.576	1886.142	1886
70	0.8384	310.272	1894.93	1895
71	0.8646	259.968	1902.4	1902
72	0.8908	209.664	1908.552	1909
73	0.917	159.36	1913.387	1913
74	0.9432	109.056	1916.903	1917
75	0.9694	58.752	1919.101	1919
76	0.9956	8.448	1919.981	1920

We will be writing software using a high-level language called PICBASIC PRO. When using code requiring a compiler, the programmer cannot be sure of the exact execution time until the code is compiled. The author's goal was to ensure that the execution time would be much less than the interrupt period. The reader is invited to try higher-resolution encoders, being careful to check the length of the assembly language code generated by the compiler, and estimating the code execution time as it is being developed. To cite one possible problem area, a lookup table is used for position feedback in the code that follows later in the chapter. The number of entries in this table would change from 76 to 308. This would result in a longer program taking more time to execute.

8.4 Hardware and Software Development

It was decided that program 15.10 of [Sandhu, 2009] would be used as a starting point for further development. A move is first preprogrammed as previously discussed, using a lookup table. The move is made, the program pauses, and the cycle is repeated indefinitely or until power is removed. A 2 sec pause was selected for a 2 sec move, so that the repeatability of the move over time could be examined. The 2 sec pause was also sufficient to detect any oscillation at the end of each move. The accuracy of each repeated move was assessed by marking the encoder wheel with a Sharpie-type marker with a thin radial line and observing the relative position of the line to some fixed reference point on the encoder reader board. The entire hardware setup is shown in Figure 8.9.

We will use the words *software* and *firmware* interchangeably, recognizing that every time a program is written to the flash program memory of the PIC18F4331, it is "firm." *Firm* means the program is located on-chip in non-volatile memory, with a typical retention life of 100 years. The beauty of the flash memory technology is that it can be erased and reprogrammed up to 100,000 times using Microchip's in-circuit serial programming (ICSP). This makes reprogramming the device very convenient, and code debugging faster than in the past.

8.4.1 Software Development

The software to be presented here was developed using the MicroCode Studio platform by Mechanique for MS Windows XP, and the PICBASIC PRO compiler written by MicroEngineering Labs, Inc. These two applications work in harmony with one another to produce efficient Intel .HEX files for the Microchip line of microcontrollers. Listing 8.1 compiled to about 2 KB of

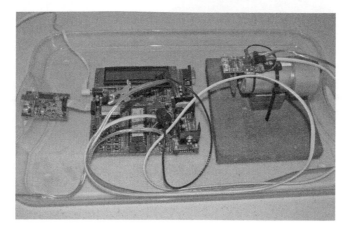

FIGURE 8.9
Entire hardware setup used for digital servo implementation. From left to right: ME Labs USB programmer; LAB-X1 board with Xavien amplifier mounted in prototype area; retrofitted motor/encoder combination mounted on V-block.

program memory and was successful in meeting the above requirements. We wished to overcome the effects of Coulomb friction. Including an error integration routine would have required signed arithmetic. Because this is not available in PICBASIC PRO, a compromise was reached which was to apply some minimum gain at zero error, in an otherwise proportional control loop. This is useful at the beginning of a move but also keeps power applied at the move end, which can cause an overshoot of the final commanded position. By immediately disabling the PWM output when the last lookup was made, this overshoot was avoided. Using the triangular velocity profile, no oscillation at the end of the move was observed. A nonlinear gain curve is remarkably simple to program in digital servo control but difficult to implement in an analog controller. Sandhu's Program 15.4 was used to determine a Coulomb friction offset. This resulted in the gain function used in Listing 8.1, which is shown graphically in Figure 8.10. The observed move distance tolerance was ±1 count in each 1920-count move, noncumulative, run continuously over a 36 hr period. This was verified by (1) observing the LEDs indicating the state of each channel on the Encoder Geek reader board at the end of each move, and (2) observing the Sharpie line drawn on the encoder wheel.

8.4.2 Notes on Implementation

The arrangement of Figure 8.11 was used to close the position loop using Listing 8.1. As previously described, an incremental encoder was added to the motor's rear shaft extension to control its position. The PIC16F877 was replaced with a PIC18F4331. The clock frequency was changed from 4 MHz to 20 MHz.

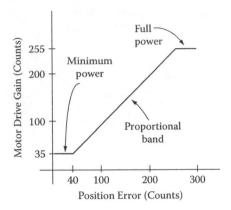

FIGURE 8.10
Nonlinear gain function used in Listing 8.1.

Because more electrons are transported per unit time at higher frequencies
(e.g., on and off the plates of a capacitor), the current draw increases with clock
frequency for the same capacitor charge. A generous heat sink was added to
the 7805 regulator on the LAB-X1 to allow the extra current to be drawn without overheating the regulator. This heat sink also acts as a convenient contact for a single-point ground connection between digital ground and power
ground. The author used a spare encoder bracket as a heat sink and connected

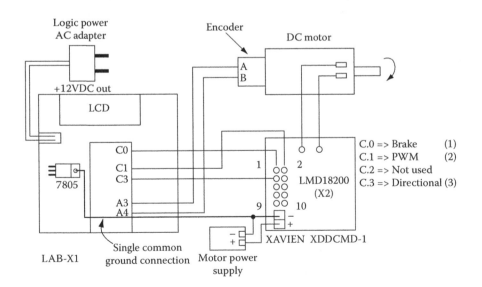

FIGURE 8.11
Digital control of a DC motor, block diagram (Figure 7.9 with power supplies and grounding
added). Processor used on the LAB-X1 is the PIC18F4331.

a Pomona-type hook lead between it and the motor power supply ground. Finally, the ME Labs USB programmer was configured with the "HS" (high-speed crystal resonator) setting instead of the default "XT" setting.

LISTING 8.1

Preprogrammed digital position control using a PIC18F4331. (Adapted from *Running Small Motors with PIC Microcontrollers* by Harprit Singh Sandhu, © 2009 by the McGraw-Hill Companies. Used by permission.)

```
'****************************************************************
'* Name : Program3 ch8 rev6.pbp *
'* Author : Stephen M. Tobin *
'* Notice : Copyright (c) 2009 Optical Tools Corporation *
'* : All Rights Reserved *
'* Date : 11/22/2009 *
'* Version : 6.0 *
'* Notes : Adapted from Program 15.10 of Sandhu *
'****************************************************************

'
CLEAR                    ; clear memory
DEFINE OSC 20       ; 20 MHz clock
DEFINE LCD_DREG PORTD        ; define lcd connections
DEFINE LCD_DBIT 4        ; 4 data bits
DEFINE LCD_BITS 4         ; data starts on bit 4
DEFINE LCD_RSREG PORTE        ; select register
DEFINE LCD_RSBIT 0       ; select bit
DEFINE LCD_EREG PORTE        ; enable register
DEFINE LCD_EBIT 1        ; select bit
LOW PORTE.2                 ; set bit low for writing to the lcd
DEFINE LCD_LINES 2        ; lines in display
DEFINE LCD_COMMANDUS 2000        ; delay in micro seconds
DEFINE LCD_DATAUS 50  ; delay in micro seconds
DEFINE ADC_BITS 8        ; set number of bits in result
DEFINE ADC_CLOCK 3      ; set clock source (3=rc)
DEFINE ADC_SAMPLEUS 50       ; set sampling time in us
DEFINE CCP2_REG PORTC        ; hpwm 2 pin port
DEFINE CCP2_BIT 1         ; hpwm 2 pin bit 1
CCP1CON = %00111111      ; set status register
TRISA = %00011111  ; set status register
```

```
LATA = %00000000 ; set status register
TRISB = %00000000 ; set status register
LATB = %00000000  ; set status register
TRISC = %00000000 ; set status register
TRISD = %00000000 ; set status register
ANSEL0 = %00000001          ; page 251 of data sheet, status register          ;
ANSEL1 = %00000000          ; page 250 of data sheet, status register          ;
QEICON = %10001000          ; page 173 counter set up, status register;
INTCON = %10101100          ; set interrupt status register
INTCON2.7 = 0               ; set status register
T0CON = %10000000           ;
POSITION VAR WORD           ; set variables
TARGET VAR WORD             ; set variables
ERROR VAR WORD              ; set variable
MOTSPD VAR WORD             ; potentiometer position
MOVDST VAR WORD             ; potentiometer position
POTVAL VAR WORD             ; potentiometer position
MOTPWR VAR WORD             ; motor power
INTNUM VAR WORD             ;
COUNTER VAR WORD            ;
X VAR WORD                  ;
INTNUM = 0;          ;
MOTPWR = 0           ;
PORTC.0 = 0          ; brake off, motor control
PORTC.1 = 0          ; PWM bit for channel 2 of hpwm
PORTC.3 = 0          ; direction bit for motor control
PORTD = 0                   ;
PAUSE 500                   ; lcd start up pause
LCDOUT $FE, $01, "START/CLEAR"          ; clear message
PAUSE 500                   ; pause to see message
GOSUB RSTPOS     ;
ON INTERRUPT GOTO INTROT  ; interrupt routine
                            ;

LOOP:                       ; main loop
     ADCIN 0 ,POTVAL    ; read pot value
     POTVAL=POTVAL/2            ; and divide it by 2
     POSITION=256*POSCNTH +POSCNTL       ; figure count
     GOSUB FIGURE_ERROR      ; and direction
```

```
        GOSUB RUNMOTOR          ; run motor
        IF ERROR=0 AND TARGET=2920 THEN  ; my final value for position
            HPWM 2, 0, 20000          ; C.1 PWM signal set to 0
                GOSUB RSTPOS       ; reset the position target
    PAUSE 2000          ; and pause
      ENDIF                 ;
GOTO LOOP           ; go back to loop
                              ;
FIGURE_ERROR                  :
     IF POSITION<TARGET THEN    ; not yet there
            ERROR=TARGET-POSITION         ; figure error
            PORTC.3=1   ; set motor direction forward
     ELSE               ;
            ERROR=POSITION-TARGET          ; figure error
            PORTC.3=0     ; set motor direction reverse
     ENDIF              ;
RETURN                 ;
                          ;
RUNMOTOR:             ;
     SELECT CASE ERROR       ; decide what to do for each INTERRUPT
            CASE IS>300;
                MOTPWR=255           ; full power
            CASE IS>40   ;
                MOTPWR=ERROR        ; proportional error
            CASE ELSE   ;
                MOTPWR=35; minimal power +10 counts to move
                Hansen gearmotor
     END SELECT ;
     IF MOTPWR>255 THEN MOTPWR=255  ; make sure value is not
     overflowing
     HPWM 2, MOTPWR, 20000           ; C.1 PWM signal
RETURN                 ;
                          ;
DISABLE               ;
INTROT:                     ; interrupt routing details
'the 1000 count offset is added to each value
  LOOKUP2 INTNUM,[1000,1001,1003,1006,1011,1016,1024,1032,1042,1053,_
  1066,1080,1095,1111,1129,1148,1169,1190,1214,1238,1264,1291,1319,_
  1349,1380,1412,1445,1480,1517,1554,1593,1633,1675,1718,1762,1807,_
```

```
1854,1902,1952,2010,2058,2105,2151,2195,2238,_
2280,2320,2359,2397,2434,2469,2503,2535,2566,2596,2625,2652,_
2678,2702,2726,2748,2768,2788,2806,2822,2838,2852,2865,2876,_
2886,2895,2902,2909,2913,2917,2919,2920],TARGET
INTNUM=INTNUM+1     ; increment counter
'have to shut off timer at end of lookup table!
IF INTNUM>76 THEN
   T0CON=%00000000 ;
   INTNUM=0
ENDIF
      INTCON.2 = 0          ; clear the interrupt bit
RESUME                     ;
ENABLE                     ;
                                    ;
RSTPOS:                    ; reset counters
   T0CON=%00000000 ;
      POSCNTH = 3          ; set counter for encoder, h bit
      POSCNTL = 232        ; set counter for encoder, l bit
      POSITION=1000        ;
      TARGET=1000          ;
      T0CON=%10000000   ;
      INTNUM=0             ;
RETURN                     ;
                                  ;
SHOW_LCD:          ; display subroutine
      LCDOUT $FE, $80, "TAR=",DEC5 TARGET, " GAIN=",DEC3 MOTPWR
      LCDOUT $FE, $C0, "POS=",DEC5 POSITION," ERROR=",DEC3 ERROR
RETURN                     ;
                                ;
END                     ; all programs must end with end
```

References

Atmel Corp., "AVR221: Discrete PID Controller," Application note, 2006.
Baron, W. et al., "Development of a High-Performance, Low-Mass, Low-Inertia Plotting Technology," *Hewlett-Packard Journal*, Oct. 1981.
Halliday, D., and R. Resnick, *Physics, Part 1*, Wiley, 1977.

Hunt, S., "Increasing the High Speed Torque of Bipolar Stepper Motors," National Semiconductor AN-828, 1993.

Kuo, B., *Digital Control Systems*, Oxford, 1995.

Kuo, B., *Incremental Motion Control: Step Motors and Control Systems*, SRL, 1978.

Leigh, J., *Applied Digital Control*, Dover, 1992.

Microchip Technology, "PIC18F2331/2431/4331/4431 Data Sheet," 2007.

Sandhu, H., *Running Small Motors with PIC Microcontrollers*, McGraw-Hill, 2009.

Valenti, C., "Implementing a PID Controller Using a PIC18 MCU," AN-937, 2004.

Appendix A: The R/C Hobby Servo

A.1 Introduction

The R/C (radio-controlled) "hobby" servo is a prepackaged analog DC servo with potentiometer feedback. It generally takes less than 10 W of input power, making it quite small, limiting its use to hobby-type applications. However, because of its ease of use and standardized drive specifications, these devices have recently become popular in the realm of education, robotics, and small commercial products requiring less than 360° of mechanical motion. The *R/C* and *hobby* designations are from the original application of these servos, the radio-controlled motion of the various control surfaces of model airplanes while in flight [Coelho, 1993]. Hobby servos are now being driven directly by wire as well as by radio control, as hobbyists and educators build small robots and other land-based machines. Figure A.1 shows a photograph of a typical R/C servo package. We will refer to these unique devices as "RC" servos (dropping the "/") or simply "servos" as we move forward to discuss the details of their construction.

A.2 Digital versus Analog Designation

Because the RC servo contains both analog and digital circuitry, it is difficult to classify. The fully analog shaft repeater described earlier in the book was invented prior to the RC servo. Manufacturers then asked themselves how to make these shaft repeaters respond to radio signals and how to shrink their size and weight to the point where two, three, or even four of these could be carried aboard a gasoline-powered model airplane. Radio signals are either amplitude or frequency modulated, making them essentially AC signals. The problem was to come up with a standard signal format that could be demodulated and then applied to the input of the servo controller inside each servo package. In this book, we have been arbitrarily classifying DC servos according to the type of feedback signal they employ. Accordingly, digital servos employ discrete data feedback from an encoder and analog

FIGURE A.1
Photograph of a typical R/C servo package, with horn installed.

servos employ continuous data feedback from a potentiometer. This classifies RC servos as analog devices, even though discrete data packets are used to drive them. These data packets (pulses) can come from either an application-specific IC like the Signetics NE544 or a small microcontroller. If the servo uses a microcontroller to drive the motor's H-bridge, manufacturers call them "digital" servos. However, a potentiometer is still used in these devices [Hobbico, 2010], and therefore we will consider them to be fundamentally analog in nature.

A.3 Anatomy of an RC Servo

Figure A.2 shows a cross-sectional view of the mechanical parts of an RC servo. The output shaft of a small DC motor is connected through a multiple-reduction gear train to the package output shaft. On the external end of the output shaft, a spline is cut, to connect various actuator arms or "horns." A single-turn potentiometer is connected to the internal end of the output shaft, to provide a position feedback signal to the electronics. Mechanical stops molded into the package prevent the output shaft from turning more than 180°. The input to the device is three-wire, consisting of V_{CC}, ground, and the command signal. V_{CC} is a nominal +5 VDC.

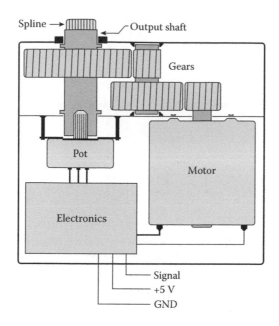

FIGURE A.2
Electromechanical schematic for an RC servo. (Reproduced with permission of Jim Stewart and *Servo Magazine*, © 2008 T&L Publications.)

A.4 Input Signal and Resulting Motion

According to [Philips/Signetics, 1988], the required input signal is a repetitive pulse whose width is between 1.0 and 2.0 ms. [Stewart, 2008] specifies that the pulses are spaced 20 ms apart, giving a repetition rate of 50 Hz. This is essentially the update or "sampling" rate of this hybrid analog/digital device. As shown in Figure A.3, a train of 2.0 ms pulses positions the servo at one end of its travel—for example, +90°. If the pulses are narrowed to 1.5 ms, the servo moves to its center or 0° position. Further narrowing the pulses to 1.0 ms brings the servo to the other end of its travel—in this case, –90°. The pulse train can be supplied to the servo in two ways. The first is direct drive, and an example of this is Stewart's "Servo Buddy" circuit. Another example of direct drive is a microcontroller output port directly connected to the input wire, with appropriate software programmed to drive the port with a PWM output. The second

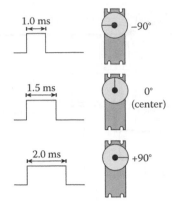

FIGURE A.3
Input pulse width versus output shaft position for an RC servo. (Reproduced with permission of Jim Stewart and *Servo Magazine,* © 2008 T&L Publications.)

way of supplying pulses to one or more servos is by radio link. The general sequence is as follows:

1. Generate pulses for each servo and encode for transmission (Signetics NE5044).
2. Modulate and transmit (Motorola MC2833).
3. Receive and demodulate (Motorola MC3361).
4. Decode transmission (Signetics NE5045).
5. Drive each servo (Signetics NE544).

ICs commonly used for each function follow in parenthesis for a typical narrow-band FM system with a range of 50 m. A full discussion of radio linking is beyond the scope of this book, but interested readers may consult [Coelho, 1993], [Foster, 1981], and the ARRL handbook [ARRL, 2001]. For the remainder of this discussion, we will assume the servo is being driven directly.

A.5 RC Servo Electronics

Figure A.4 shows a block diagram of an RC servo. All components to the left of the motor, gears, and potentiometer are part of the inner contents of the servo, meaning that an IC such as the NE544 is housed inside the servo case. The local pulse generator is a one-shot triggered by the command input pulse. The local pulse width indicates the current position of the feedback

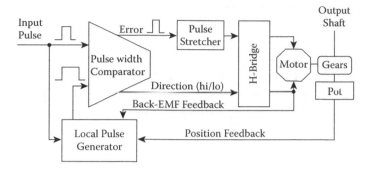

FIGURE A.4

Functional block diagram of an RC servo. (Reproduced with permission of Jim Stewart and *Servo Magazine*, © 2008 T&L Publications.)

potentiometer. The input pulse and the local pulse are fed to a pulse width comparator, whose two outputs give (1) the difference between the two pulses (error magnitude), and (2) which pulse was wider (error direction). The error magnitude pulse is then sent to a pulse stretcher, which acts as an amplifier. Both the amplified error signal and the direction signal are sent to the H-bridge to drive the motor. The process is repeated 50 times per second until the error is reduced to an acceptable tolerance, known as the dead-band. This dead-band helps compensate for Coulomb friction in the gear train. Once the error is corrected to within this tolerance, the next command pulse may ask for no additional movement. In this case, the servo will not rotate and will remain where it was last positioned. If the next command pulse asks for a change in position, the servo will respond as needed. In either case, the stream of input pulses is continuous. The NE544 designers also provided adjustable damping control. Between input pulses, the back-emf signal from the motor itself can modify the length of the next local pulse. Because the back-emf signal is proportional to the motor speed, it acts as a damper to prevent oscillation at end of travel [Kuhnle, 2008].

A.6 Circuit Example Using the NE544

A typical application circuit using the Signetics NE544 IC is shown in Figure A.5. This circuit is shown in the Signetics data sheet, but because of the age of the chip, applications information is hard to come by. The following description is quoted from [Buse, 2000]:

> A positive input signal applied to the input pin 4 sets the input flip-flop and starts the one-shot time period. The directional logic compares the length of the input pulse to that of the internal one shot and stores the

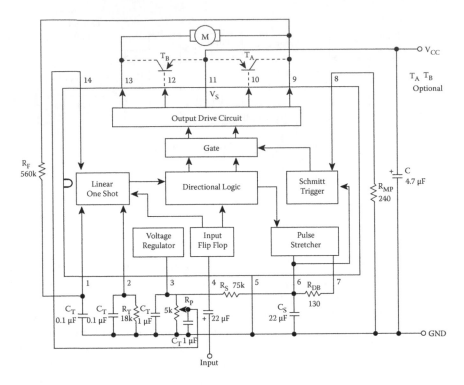

FIGURE A.5

Electrical block diagram and recommended external circuit connections for the Signetics NE544 Servo Amplifier integrated circuit.

result of this comparison (called the error pulse) and also feeds this pulse to a pulse stretcher, deadband, and trigger circuit. These circuits determine three important parameters:

1. Deadband—The minimum difference between input pulse and internally generated pulse to turn on the output.
2. Minimum output pulse—The smallest output pulse that can be generated from the Schmitt trigger circuit.
3. Pulse stretcher gain—The relationship between error pulse and output pulse.

Adjustment of these parameters is achieved with external resistors and capacitors at pins 6, 7, and 8. Deadband is controlled by resistor R_{DB}. The minimum output pulse is controlled by R_{MP}. The pulse stretcher gain is adjusted by capacitor C_S and resistor R_S. The trigger circuit activates the gate for a precise length of time to provide drive to the bridge output circuitry in proportion to the length of the error pulse. Resistor R_F determines the amount of feedback required for good closed loop damping.

T_A and T_B are external PNP transistors for increased motor drive, which make a faster, more powerful servo with better resolution. The amount of servo travel is controlled by resistor R_T and can be varied to change the amount of servo rotation. Increase the value of R_T to decrease servo travel or decrease the value of R_T to increase servo travel.

This information is still valuable because virtually all of the smaller, more capable custom ICs used in RC servos today are based on the same concepts.

References

American Radio Relay League, *The ARRL Handbook*, 78th ed., 2001.

Buse, L., "Radio Control Servos and Speed Control" (quote previously attributed to Ace R/C), http://www.seattlerobotics.org/encoder/200009/Servos.html, August 2000.

Coelho, A., "Understanding Your Radio," http://www.rchelibase.com/radio/radio. pdf, 1993.

Foster, C., *Real Time Programming—Neglected Topics*, Addison-Wesley, 1981.

Hobbico, Inc., "Futaba Digital FET Servos," http://www.futaba-rc.com/servos/digitalservos.pdf, 2010.

Kuhnle, W., "Bio-Feedback: Letter to the Editor," *Servo*, July 2008.

Philips/Signetics Corp., "NE544 Servo Amplifier Product Specification," 1988.

Stewart, J., "The Servo Buddy," *Servo*, May 2008.

Bibliography

American Radio Relay League, *The ARRL Handbook*, 78th ed., 2001.

Atmel Corp., "AVR221: Discrete PID Controller," Application note, 2006.

Barello, L., "H-Bridges and PMDC Motor Control Demystified," http://www.barello.net/papers/index.htm, 2003.

Baron, W. et al., "Development of a High-Performance, Low-Mass, Low-Inertia Plotting Technology," *Hewlett-Packard Journal*, Oct. 1981.

Black, H., "Stabilized Feedback Amplifiers," *Bell Sys. Tech. J.*, v. 13, 1934.

Bode, H., *Network Analysis and Feedback Amplifier Design*, Van Nostrand, 1945.

Brown, G., and D. Campbell, *Principles of Servomechanisms*, Wiley, 1938.

Buse, L., "Radio Control Servos and Speed Control" (quote previously attributed to Ace R/C), http://www.seattlerobotics.org/encoder/200009/Servos.html, August 2000.

Cannon, R., *Dynamics of Physical Systems*, McGraw-Hill, 1967.

Chapman, S., *Electric Machinery Fundamentals*, WCB McGraw-Hill, 1999.

Coelho, A., "Understanding Your Radio," http://www.rchelibase.com/radio/radio.pdf, 1993.

DiStefano, J. et al., *Feedback and Control Systems*, Schaum, 1967.

Doebelin, E., *Control System Principles and Design*, Wiley, 1985.

Doebelin, E., *Dynamic Analysis and Feedback Control*, McGraw-Hill, 1962.

Doebelin, E., *System Dynamics: Modeling and Response*, Merrill, 1972.

Electro-Craft Corp., *DC Motors, Speed Controls, Servo Systems*, 5th ed., 1980.

Ellement, D., and M. Majette, "Low-Cost Servo Design," *Hewlett-Packard Journal*, Aug. 1988.

Epstein, H. et al., "An Incremental Optical Shaft Encoder Kit with Integrated Optoelectronics," *Hewlett-Packard Journal*, Oct. 1981.

Epstein, H. et al., "Economical, High-Performance Optical Encoders," *Hewlett-Packard Journal*, Oct. 1988.

Evans, W., *Control System Dynamics*, McGraw-Hill, 1954.

Foster, C., *Real Time Programming—Neglected Topics*, Addison-Wesley, 1981.

Franco, S., *Design with Operational Amplifiers and Analog Integrated Circuits*, McGraw-Hill, 1988.

Fredricksen, T., *Intuitive Operational Amplifiers*, McGraw-Hill, 1988.

Friedland, B., *Control System Design: An Introduction to State-Space Methods*, Dover, 1986.

Gardner, M., and J. Barnes, *Transients in Linear Systems*, Wiley, 1956.

Halliday, D., and R. Resnick, *Physics, Part 1*, Wiley, 1977.

Harrison, P., "Quadrature Decoding with a Tiny AVR," http://www.helicron.net/avr/quadrature, 2005.

Hayes, T., and P. Horowitz, *Learning the Art of Electronics*, Cambridge, 2010.

Hobbico, Inc., "Futaba Digital FET Servos," http://www.futaba-rc.com/servos/digitalservos.pdf, 2010.

Hunt, S., "Increasing the High Speed Torque of Bipolar Stepper Motors," National Semiconductor AN-828, 1993.

Jackson, L. et al., "Deskjet Printer Chassis and Mechanism Design," *Hewlett-Packard Journal*, Oct. 1988.

Kuhnle, W., "Bio-Feedback: Letter to the Editor," *Servo*, July 2008.

Kuo, B., and J. Tal, *Incremental Motion Control: DC Motors and Control Systems*, SRL, 1978.

Kuo, B., *Automatic Control Systems*, Prentice-Hall, 1982.

Kuo, B., *Digital Control Systems*, Oxford, 1995.

Kuo, B., *Incremental Motion Control: Step Motors and Control Systems*, SRL, 1978.

Leigh, J., *Applied Digital Control*, Dover, 1992.

Mancini, R., *Op Amps for Everyone*, Newnes, 2003.

MathWorks, The, "Getting Started with the Control System Toolbox," 2006.

Microchip Technology, "PIC18F2331/2431/4331/4431 Data Sheet," 2007.

National Semiconductor, "An Applications Guide for Op Amps," AN-20, 1980.

Nise, N., *Control Systems Engineering*, Wiley, 2004.

Ogata, K., *Matlab for Control Engineers*, Pearson Prentice-Hall, 2008.

Philips/Signetics Corp., "NE544 Servo Amplifier Product Specification," 1988.

Pressman, A., *Switching Power Supply Design*, McGraw-Hill, 1991.

Regan, T., "A DMOS 3A, 55V, H-Bridge: The LMD 18200," National Semiconductor AN-694, 1990.

Reswick, J., and C. Taft, *Introduction to Dynamic Systems*, Prentice-Hall, 1967.

Sandhu, H., *Running Small Motors with PIC Microcontrollers*, McGraw-Hill, 2009.

Stewart, J., "The Servo Buddy," *Servo*, May 2008.

Stiffler, K., *Design with Microprocessors for Mechanical Engineers*, McGraw-Hill, 1992.

Taft, C., and E. Slate, "Pulsewidth Modulated DC Control," *IEEE Trans. Ind. Elec. and Control Inst.*, vol. IECI-26, Nov. 1979.

Tietze, U., and Ch. Schenk, *Electronic Circuits: Design and Applications*, Springer-Verlag, 1991.

Valenti, C., "Implementing a PID Controller Using a PIC18 MCU," AN-937, 2004.

Widlar, R., "Monolithic Op Amp—The Universal Linear Component," National Semiconductor AN-4, 1968.

Index